AF347523

Impacts of Climate on Renewable Groundwater Resources and/or Stream-Aquifer Interactions

Impacts of Climate on Renewable Groundwater Resources and/or Stream-Aquifer Interactions

Editors

Francisco Javier Alcalá
David Pulido-Velázquez
Luis Ribeiro

MDPI • Basel • Beijing • Wuhan • Barcelona • Belgrade • Manchester • Tokyo • Cluj • Tianjin

Editors
Francisco Javier Alcalá
Geological Survey of Spain
Spain

David Pulido-Velázquez
Geological Survey of Spain
Spain

Luis Ribeiro
University of Lisbon
Portugal

Editorial Office
MDPI
St. Alban-Anlage 66
4052 Basel, Switzerland

This is a reprint of articles from the Special Issue published online in the open access journal *Water* (ISSN 2073-4441) (available at: https://www.mdpi.com/journal/water/special_issues/renewable_groundwater).

For citation purposes, cite each article independently as indicated on the article page online and as indicated below:

LastName, A.A.; LastName, B.B.; LastName, C.C. Article Title. *Journal Name* **Year**, *Volume Number*, Page Range.

ISBN 978-3-0365-1357-7 (Hbk)
ISBN 978-3-0365-1358-4 (PDF)

Contents

About the Editors

Francisco Javier Alcalá is Senior Researcher at the Geological Survey of Spain. His research interest deals with human and global interactions underlying problems of groundwater quantity and quality for strategies of diagnosis, operation, mitigation, and protection. Innovations deals with applications for groundwater recharge and chemical baseline evaluations at different spatiotemporal scales and climate scenarios. He has experience as Principal Investigator in National and International R&D projects. He is author of 49 high-impact SCI papers, and has significant activity as reviewer, editor, and representation in international committees and forums.

David Pulido-Velázquez is Senior Researcher at the Geological Survey of Spain. His research focus on the development of methodologies and tools for integrated analysis of management alternatives in water scarcity systems within the Mediterranean area, with special emphasis on the study of impacts and adaptation strategies to global change. He has experience as Principal Investigator in National and European R&D projects. He is author of 47 high-impact SCI papers and has significant activity as reviewer and editor.

Luis Ribeiro was Full Professor of Groundwater Hydrology at the Instituto Superior Técnico from University of Lisbon. His research interests focused on groundwater numerical and stochastic models, groundwater and global change, groundwater dependent ecosystems, integrated management of water resources, and water in ancestral civilizations. Prof. Ribeiro passed away on 3 April 2020. Her devotion to groundwater research leaves a legacy of multidisciplinary knowledge to the scientific community and to the public in general.

Preface to "Impacts of Climate on Renewable Groundwater Resources and/or Stream-Aquifer Interactions"

This book collects recent and original contributions in the field of climate and underlying human influences on renewable groundwater resources and/or stream–aquifer interactions. The readers can find the contributions both interesting and inspiring when exploring the field of impacts of climate (past, current, and future) on natural, premeditated, and unpremeditated aquifer recharge sources determining the renewable fraction of the groundwater resource. The findings and methods presented in these original contributions will be of interest in some associated problematics concerning stream–aquifer interactions in drylands and mountainous areas, the resilience of groundwater-dependent ecosystems to future climate change and human action, land-subsidence problems in the future, and the influence of climate change on groundwater resources availability and its implication for sustainable groundwater management policies ahead. The Editors envision that these contributions would also be of interest to researchers and practitioners and help identify further research routes.

Francisco Javier Alcalá, David Pulido-Velázquez, Luis Ribeiro
Editors

Editorial

Impacts of Climate on Renewable Groundwater Resources and/or Stream–Aquifer Interactions

Francisco J. Alcalá [1],*, David Pulido-Velazquez [2] and Luis Ribeiro [3],†

[1] Instituto Geológico y Minero de España, Ríos Rosas, 23, 28003 Madrid, Spain
[2] Instituto Geológico y Minero de España, Urb. Alcázar del Genil, 4. Edificio Zulema, Bajo, 18006 Granada, Spain; d.pulido@igme.es
[3] Civil Engineering Research and Innovation for Sustainability (CERIS), Instituto Superior Técnico, Universidade de Lisboa, 1049-001 Lisbon, Portugal
* Correspondence: fj.alcala@igme.es; Tel.: +34-913-495-840
† Deceased.

Received: 3 December 2020; Accepted: 9 December 2020; Published: 10 December 2020

1. Introduction

The evaluation of aquifer recharge is essential to make a quantitative evaluation of renewable groundwater resources required to implement proper water policies aimed at maintaining stream–aquifer interactions, guaranteeing water supply to human activities, and preserving groundwater-dependent ecosystems at different spatial and temporal scales and climate conditions. As another hydrological variable, aquifer recharge is intrinsically uncertain and may integrate different water fluxes with different hydrological origins. Aquifer recharge may include (i) natural sources from precipitation, aquifers transference, river losses, and snow melting; (ii) unpremeditated human-induced sources from irrigation and urban returns and losing channels and reservoirs; and (iii) premeditated human-routed sources from artificial infiltration practices and returns derived from non-conventional water sources such as wastewater reuse and desalination. Weather-land attributes, human-water requirements, and global-climatic forces determine the magnitude of natural sources, the existence of additional human-induced sources when water is used, and the need for human-routed sources in contexts of water scarcity. In short, these drivers determine the renewable faction of the groundwater resource in a given groundwater body.

However, a temporal perspective to understand how climate determines the aquifer recharge sources and the groundwater renewability is needed because resources on many groundwater bodies may be variably related to natural aquifer recharge produced during past climates. Therefore, for similar large groundwater storage, the influence of climate may extend over the last millennium in drylands with very low recharge rates and over the last centuries and decades in temperate mid-terrestrial-latitude semiarid and sub-humid regions, respectively. Current global climatic forces, which include the increasing influence of floods and droughts in different terrestrial latitudes and the variable human influence, condition water resources management policies in the near future. Finally, global climate scenarios foresee the thresholds for water availability ahead.

In this wide "aquifer-recharge–climate" framework, this Special Issue was aimed at reporting all influences of climate (past, current, and future) on natural, unpremeditated, and premeditated aquifer recharge sources over different aquifer and landscape typologies at different spatial and temporal scales.

2. Contributions

Since the call for papers was announced in March 2019, and after a rigorous peer-review process, six papers have been accepted for publication [1–6]. To gain a better insight into the essence of the Special Issue, we offer brief highlights of the published papers below.

The paper "Potential Impacts of Future Climate Change Scenarios on Ground Subsidence" [1] develops a new method (a parsimonious approach) to assess the impact of climate change scenarios on land subsidence related to groundwater level depletion in detrital aquifers. The Vega de Granada aquifer in southern Spain was the case study chosen. Historical subsidence was estimated using remote sensing techniques, whereas local climate change scenarios for the future horizon (2016–2045) were generated by applying a bias-correction approach. The method will allow for anticipating sustainable adaptation strategies for land subsidence in vulnerable detrital aquifer areas during critical drought periods to be assessed.

The paper "Numerical Approaches for Estimating Daily River Leakage from Arid Ephemeral Streams" [2] studies the patterns of river infiltration and the associated controlling factors in an approximately 150-km section of the Donghe River (lower Heihe River, China) by using a combination of field investigations and modeling techniques. In this arid region, the simulated infiltration was most sensitive to the aquifer hydraulic conductivity and the maximum evapotranspiration rate. Both hydraulic parameters of riverbeds and evapotranspiration parameters were equally important for quantifying the flux exchange between arid ephemeral streams and underlying aquifers. Findings are of interest to help maintain riparian ecosystems in arid regions.

The paper "Coupling SWAT Model and CMB Method for Modeling of High-Permeability Bedrock Basins Receiving Interbasin Groundwater Flow" [3] couples the Soil and Water Assessment Tool model and the chloride mass balance (CMB) method to improve the modeling of streamflow in high-permeability bedrock basins receiving interbasin groundwater flow (IGF), i.e., the naturally occurring groundwater flow beneath a topographic divide contributing to the streamflow. The Castril River basin in southern Spain was the case study chosen. In this headwater area, which has null groundwater exploitation, the mean yearly IGF was about 0.5-fold of the mean yearly baseflow. This paper provides a way to identify IGF in high-permeability bedrock basins.

The paper "The Ecosystem Resilience Concept Applied to Hydrogeological Systems: A General Approach" [4] discusses the role of resilience of hydrogeological systems affected by either climate and/or anthropic actions in order to understand how anticipating negative changes (transitions) and preserving its services. The paper reports typical human actions modifying groundwater dynamics of hydrogeological systems in recent decades, which can be increased by climate change with delayed effects on groundwater quantity and quality as rivers that have dried up, wetlands that have disappeared, leaving their buckets converted into farmland, and aquifers that have been intensively exploited for years.

The paper "Climate-Dependent Groundwater Discharge on Semi-Arid Inland Ephemeral Wetlands: Lessons from Holocene Sediments of Lagunas Reales in Central Spain" [5] analyzes the water balance of wetlands, which are systems highly sensitive to climate change and human action. The Lagunas Reales in central Spain, a semiarid inland wetland endangered by both climate and human drivers, was the case study chosen. Studying the Holocene sedimentary record found that past arid periods produced greater surface freshwater inflow and low contribution of deep saline groundwater. During past wet periods, deep saline groundwater discharge increased wetland water salinity. The paper analyzes the wetlands resilience to natural and human-induced changes.

The paper "Using the Turnover Time Index to Identify Potential Strategic Groundwater Resources to Manage Droughts within Continental Spain" [6] identifies groundwater (GW) bodies with low vulnerability to pumping to be used as potential buffer values for sustainable conjunctive use management during droughts. In each GW body, GW vulnerability was obtained using the natural turnover time index as the storage capacity divided by recharge. For the historical period and near future (until 2045), this approach was applied to the 146 Spanish GW bodies at risk of not achieving a good quantitative status, according to the target of the European Water Framework Directive. The paper contributes to sustainable adaptation strategies to adapt to climate change.

3. Conclusions

The Guest Editors envision that published papers in this Special Issue would be of interest to researchers and practitioners and help identify further research routes. We also hope that the readers can find the material of this Special Issue both interesting and inspiring when exploring the field of impacts of climate (past, current, and future) on natural, premeditated, and unpremeditated aquifer recharge sources determining the renewable fraction of the groundwater resource. The findings and methods presented in this collection of papers contribute to the increasing interest in some associated problematics concerning stream–aquifer interactions in drylands and mountainous areas, the resilience of groundwater-dependent ecosystems to future climate change and human action, land-subsidence problems in the future, and the influence of climate change on groundwater resources availability and its implication for sustainable groundwater management policies ahead.

Funding: This research received no external funding.

Acknowledgments: This Special Issue is dedicated to the memory of Luis Ribeiro, whose devotion to groundwater research leaves a legacy of multidisciplinary knowledge to the scientific community and to the public in general. The authors of this paper, who served as the Guest Editors of this Special Issue, wish to thank the journal editors, all authors submitting papers, and the referees who contributed to revise and improve the six published papers.

Conflicts of Interest: The authors declare no conflict of interest.

References

1. Collados-Lara, A.J.; Pulido-Velazquez, D.; Mateos, R.M.; Ezquerro, P. Potential Impacts of Future Climate Change Scenarios on Ground Subsidence. *Water* **2020**, *12*, 219. [CrossRef]
2. Min, L.; Vasilevskiy, P.Y.; Wang, P.; Pozdniakov, S.P.; Yu, J. Numerical Approaches for Estimating Daily River Leakage from Arid Ephemeral Streams. *Water* **2020**, *12*, 499. [CrossRef]
3. Senent-Aparicio, J.; Alcalá, F.J.; Liu, S.; Jimeno-Sáez, P. Coupling SWAT Model and CMB Method for Modeling of High-Permeability Bedrock Basins Receiving Interbasin Groundwater Flow. *Water* **2020**, *12*, 657. [CrossRef]
4. De la Hera-Portillo, A.; López-Gutiérrez, J.; Zorrilla-Miras, P.; Mayor, B.; López-Gunn, E. The Ecosystem Resilience Concept Applied to Hydrogeological Systems: A General Approach. *Water* **2020**, *12*, 1824. [CrossRef]
5. Mediavilla, R.; Santisteban, J.I.; López-Cilla, I.; Galán de Frutos, L.; De la Hera-Portillo, A. Climate-Dependent Groundwater Discharge on Semi-Arid Inland Ephemeral Wetlands: Lessons from Holocene Sediments of Lagunas Reales in Central Spain. *Water* **2020**, *12*, 1911. [CrossRef]
6. Pulido-Velazquez, D.; Romero, J.; Collados-Lara, A.J.; Alcalá, F.J.; Fernández-Chacón, F.; Baena-Ruiz, L. Using the Turnover Time Index to Identify Potential Strategic Groundwater Resources to Manage Droughts within Continental Spain. *Water* **2020**, *12*, 3281. [CrossRef]

Article

Potential Impacts of Future Climate Change Scenarios on Ground Subsidence

Antonio-Juan Collados-Lara [1,*], David Pulido-Velazquez [1], Rosa María Mateos [2] and Pablo Ezquerro [3]

[1] Department of Research on Geological Resources, Geological Survey of Spain (IGME), Urb. Alcázar del Genil, 4-Edif. Bajo, 18006 Granada, Spain; d.pulido@igme.es
[2] Geohazard InSAR Laboratory and Modelling Group, Geological Survey of Spain (IGME), Urb. Alcázar del Genil, 4-Edif. Bajo, 18006 Granada, Spain; rm.mateos@igme.es
[3] Geohazard InSAR Laboratory and Modelling Group, Geological Survey of Spain (IGME), Ríos Rosas, 23, 28003 Madrid, Spain; p.ezquerro@igme.es
* Correspondence: ajcollados@gmail.com

Received: 12 December 2019; Accepted: 8 January 2020; Published: 13 January 2020

Abstract: In this work, we developed a new method to assess the impact of climate change (CC) scenarios on land subsidence related to groundwater level depletion in detrital aquifers. The main goal of this work was to propose a parsimonious approach that could be applied for any case study. We also evaluated the methodology in a case study, the Vega de Granada aquifer (southern Spain). Historical subsidence rates were estimated using remote sensing techniques (differential interferometric synthetic aperture radar, DInSAR). Local CC scenarios were generated by applying a bias correction approach. An equifeasible ensemble of the generated projections from different climatic models was also proposed. A simple water balance approach was applied to assess CC impacts on lumped global drawdowns due to future potential rainfall recharge and pumping. CC impacts were propagated to drawdowns within piezometers by applying the global delta change observed with the lumped assessment. Regression models were employed to estimate the impacts of these drawdowns in terms of land subsidence, as well as to analyze the influence of the fine-grained material in the aquifer. The results showed that a more linear behavior was observed for the cases with lower percentage of fine-grained material. The mean increase of the maximum subsidence rates in the considered wells for the future horizon (2016–2045) and the Representative Concentration Pathway (RCP) scenario 8.5 was 54%. The main advantage of the proposed method is its applicability in cases with limited information. It is also appropriate for the study of wide areas to identify potential hot spots where more exhaustive analyses should be performed. The method will allow sustainable adaptation strategies in vulnerable areas during drought-critical periods to be assessed.

Keywords: ground subsidence; climate change; Vega de Granada aquifer

1. Introduction

In many agricultural regions, as well as areas with a rapid urbanization and population growth, prolonged groundwater exploitation due to increasing water demand is causing land subsidence impacts [1–7]. In many coastal and delta cities, such as Ho Chi Minh, Bangkok, Manila, Tokyo, and Jakarta, land subsidence considerably exceeds (by up to 10 times) absolute sea level rise, which increases flood vulnerability and triggers severe, damaging impacts [8]. Sinking cities—within the framework of global change—is and will be a transnational threat.

Land subsidence is related to falling groundwater levels in (generally) unconsolidated alluvial or basin-fill aquifers with a significant proportion of compressible fine-grained materials. Increases in effective stresses—caused by headwater declines—determine the aquifer system compaction at local or regional scales [9]. This deformation is typically elastic (reversible) and results in small vertical displacements, but when the aquifer is subjected to head declines that exceed the critical levels, much of the compaction is related to an inelastic deformation and the accompanying subsidence is permanent [10].

Differential interferometric synthetic aperture radar (DInSAR) techniques are satellite-based remote-sensing tools that have been applied successfully for monitoring land subsidence thanks to their high spatial and temporal coverage, fast data acquisition, and low cost [11]. They are capable of measuring mm-scale ground displacements at a spatial resolution of 5–10 m over large regions (hundreds to thousands of square kilometers). They have been widely applied in numerous recent studies of land subsidence triggered by intense groundwater withdrawals [12–15]. A prominent case is Jakarta, the capital city of Indonesia (around 10 million people), where subsidence values of 28 cm/year have been registered in some locations. The impacts in Jakarta have been seen in several forms: not only cracking and damage in buildings and infrastructure, but subsidence also enlarges the (tidal) flooding inundation areas and makes the coast more vulnerable to sea-level-rise phenomena [1]. Indonesian authorities think the capital should be relocated.

In the present work, we applied the persistent scatterer interferometry technique (PSInSAR), an operational tool for precise ground deformation mapping, acting as a geodetic network [16]. This technique allows quantification of deformation measurements, combining them with geological and hydrogeological data in a geographical information system (GIS). The application of PSInSAR has improved rapidly in the last decade, and it is now a valuable tool for monitoring seasonal and long-term aquifer-system responses to groundwater pumping.

In the literature, we found many different approaches that have been employed to model the impact of groundwater level depletion on the subsidence. Some of them are based on physically based distributed hydro-geomechanical models [17,18], but we also found other approaches such as machine learning [19] and conceptual [20] or simple regression approaches [21], like the one that was employed in this work.

In order to assess potential future impacts of climate change (CC) in a rational way, we need to use the climatic scenarios generated by simulating with climatic models the emission scenarios identified by the Intergovernmental Panel of Climate Change (IPCC). The most recent of these are those included in the 5th assessment report [22], the RCP scenarios. These scenarios are not predictions of future climate; they are internally consistent pictures of plausible future climates that constitute a basis for other workers to evaluate the possible impacts of CC [23]. In order to make this information relevant to assessment of impacts in specific case studies, they have to be translated to regional and local scales by applying statistical correction techniques that take into account historical data [24].

Despite the spread of uncertainties involved in the assessment of future CC impacts, there is no excuse for delays or inaction in assessing/identifying adaptation strategies, taking into account that there are environments that could be very vulnerable [25]. The market for technologies for adaptation to CC is growing rapidly, given that "the cost of repairing damages is estimated to be six times greater than adaptation costs" (H2020WATER-2014/2015). In the literature, we found multiple examples of the propagation of future CC scenarios in order to assess hydrological impacts at different scales, including continental [26,27], country [28,29], river basin [30,31], and aquifer [28,32] systems. Nevertheless, the literature on the assessment of the potential future impacts of CC on land subsidence is very limited [33]. Only a few works have been developed on identification and assessment of potential adaptation strategies, like the work published by Brouns et al. [34], in which a bottom-up approach was applied to define the scenarios.

The main objective of this research was to propose a new, parsimonious approach [35,36] with which to perform a first assessment of potential CC impacts on land subsidence related to groundwater-level depletion in detrital aquifers. We also applied and validated it in the Vega de Granada aquifer (southern Spain). We have proposed a general approach applicable to any type of aquifer including in coastal areas, which are more vulnerable to subsidence [37,38]. The main innovative aspect of the proposed methodology is its applicability in cases with limited information through a combination of several parsimonious techniques: (1) a statistical approach for CC generation, (2) a simple hydrological approach to assess the impacts of CC on groundwater levels, and (3) simple linear regression models to propagate the impacts to land subsidence. The model might be also useful for a preliminary assessment of different adaptation strategies. It does not require a distributed groundwater flow model of the system [39,40]. Taking into account the uncertainties around future potential CC scenarios, the model provides a useful first approach, even in wide areas (e.g., country or continental scale), that can identify potential hot spots where more exhaustive analyses should be performed.

2. Case Study

The Vega de Granada is a flat region located in the metropolitan area of the city of Granada (southern Spain, 530,000 people) (see Figure 1). With an extension of 200 km^2, the Vega de Granada is a traditional agricultural region with increasing urban growth during recent decades. Both agricultural and urban demands exert a high pressure on the aquifer. The river Genil flows (from SE to NW) through the center of the basin, being the aquifer's main drainage axis [41].

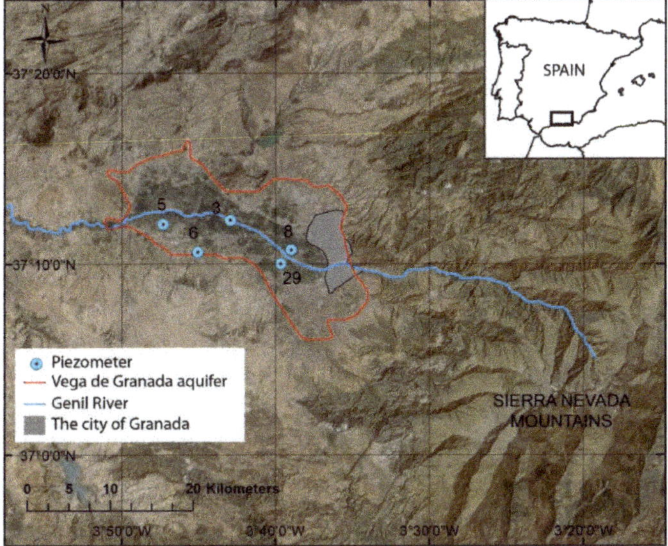

Figure 1. Location of the case study. The Vega de Granada aquifer is of regional importance. It is located within the Granada Basin domain (southern Spain) and in the central sector of the Betic cordillera (Sierra Nevada).

From the geological point of view, the Vega de Granada is within the Granada Basin domain, in the central sector of the Betic cordillera. It is a tectonic depression delineated by normal faults which control the basin infill. The Vega de Granada detrital aquifer presents a multilayer and heterogeneous structure with quaternary levels of gravel, sand, silt, and clay, with a maximum thickness of 250 m in its central part [15,42,43]. Both the river sedimentation and the surroundings alluvial fans (related to the normal faults) determine the facies distribution.

The Vega de Granada aquifer is of regional importance. With renewable water resources of about 160 hm^3/yr [43], groundwater exploitation has intensified considerably in recent decades because of urban sprawl, particularly during the severe droughts that periodically affect the region. The heterogeneous sedimentary structure controls the great spatial variability of the hydraulic parameters in the aquifer (permeability, transmissivity, etc.) and the distribution of the most productive wells. All these variables in the aquifer can be used to explain its spatial and temporal response to hydraulic head changes and the subsequent vertical ground movements.

3. Data and Methods

A flowchart of the proposed method has been represented in Figure 2.

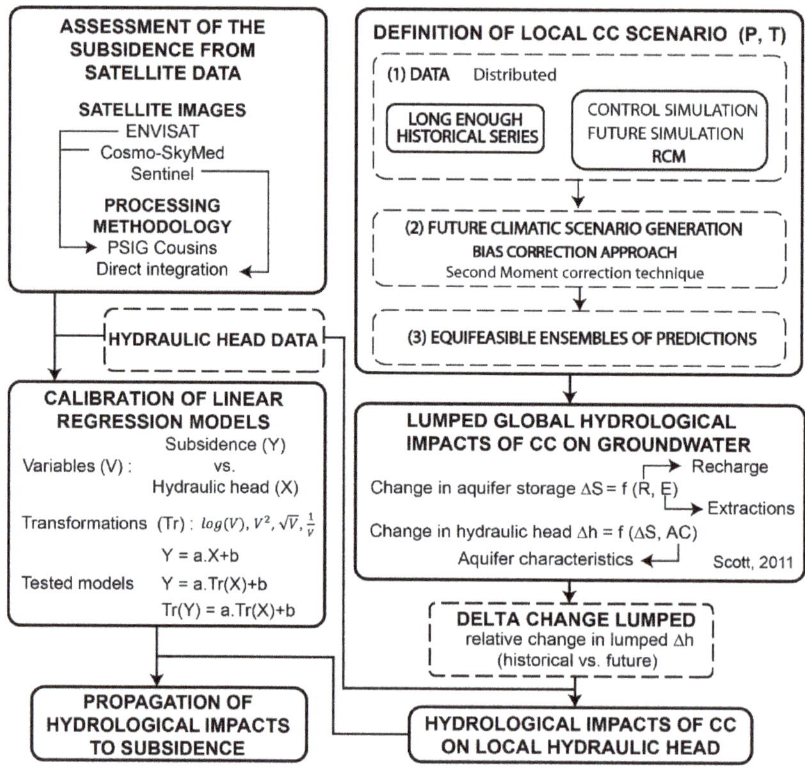

Figure 2. Flow chart of the proposed methodology.

3.1. Data Employed and Their Origins

The historical climate data for our case study were taken from Version 04 of the Spain02 dataset [44] (see Figure 3a,b). The Spain02 dataset includes daily temperature and precipitation estimates from observations (around 2500 quality-control stations) of the Spanish Meteorological Agency.

We also used data from individual global future climate projections (see Table 1) produced by regional climate models (RCM) nested within different global circulation models (GCM). This information was retrieved from the Coordinated Regional Downscaling Experiment (CORDEX) project [45]. The Spain02 project uses the same grids as the EURO-CORDEX project, which has a spatial resolution of approximately 12.5 km.

In addition, we also employed some hydrological information, including hydraulic head evolution in different piezometers obtained from the Spanish official network for monitoring the quantitative state of groundwater (see their location in Figure 1) and the historical recharge and pumping rates within the aquifer, which were taken from the information included in the Guadalquivir River Basin Plan (2015–2021).

Table 1. Regional climate models (RCMs) and global circulation models (GCMs) considered.

RCM \ GCM	CNRM-CM5	EC-EARTH	MPI-ESM-LR	IPSL-CM5A-MR
CCLM4-8-17	×	×	×	
RCA4	×	×	×	
HIRHAM5		×		
RACMO22E		×		
WRF331F				×

For those wells, we also used information about the land subsidence rates obtained by applying PInSAR techniques. The average vertical displacement based on the time series was estimated by using buffer areas with a radius of 1000 m for each piezometer and considering all PSI (persistent scatterer interferometry) data included in the area [15]. The percentage of fine-grained material (clay and silt) for each piezometer was obtained by exploiting geological data recorded in 38 boreholes drilled in the area. They were interpreted [15] to provide isolines of fine-grained sediment percentage, taking into account the clay and silt content in the first 50 m of the borehole, as the borehole depths were quite variable (from 50 to 122 m).

The results showed higher clay content in the piezometers located in the northern and southern extremes of the aquifer, as well as in its central part (see Table 2).

Table 2. Content of clay and silt in the considered piezometers.

Piezometer	Clay and Silt Content (%)
5	40
3	5
6	70
8	20
29	20

The historical data and the future trends of population within the area were obtained from the Spanish Statistical Office <https://www.ine.es/>.

3.2. Assessment of Subsidence from Satellite Data

In the present study, spatial and temporal ground surface deformation assessment was conducted by exploiting 70 SAR images from three different satellites: ENVISAT (2003–2009, C band), Cosmo-SkyMed (2011–2014, X band), and Sentinel 1A (2015–2016, C band). The processing methodologies applied for each dataset were: (1) PSIG Cousins [46] for the ENVISAT and COSMO-Sky-Med datasets, and (2) direct integration approach [47] for the Sentinel 1A dataset. More detailed specifications can be found in Mateos et al. [15]. This combination allowed a thorough assessment of the ground deformation pattern in the aquifer and both the temporal and spatial dimensions of the subsidence.

PSInSAR measurements were obtained in the aquifer of the Vega de Granada (southern Spain), covering a large temporal span of 13 years (from 2003 to 2016). In this time, a severe drought affected the area during the ENVISAT period (2003–2006), and greater groundwater withdrawals took place.

PSInSAR data were correlated (temporally and spatially) with hydraulic head changes in the aquifer along the monitoring period, and with geological data (from boreholes) regarding the clay and silt content in the aquifer. Based on the borehole information, isolines of fine sediments percentage were obtained by Mateos et al. [15]. Clay and silt content is key information which can explain the spatial response of and aquifer system to hydraulic head changes and the subsequent vertical land movements.

3.3. Definition of Local CC Scenarios

A statistical method was used to define local future global change scenarios for the pilot, based on the historical information for the adopted reference period (1986–2015) and the available RCM simulations.

These scenarios were derived from RCM simulations available in the CORDEX project [45] for the most pessimistic emission scenario, RCP 8.5, and the temporal horizon 2016–2045. The future series were generated by applying the first- and second-moment correction technique under the bias-correction approach [24]. The bias-correction techniques applied a perturbation (transformation function) to the control series of the RCM simulations to obtain another series with statistics more similar to the historical series. The transformation function in the first- and second-moment correction technique is defined by focusing on the mean and standard deviation of the climate series. The same transformation function was applied to the future simulations of the RCM to obtain the climate change projections. An equifeasible ensemble of the individual climate change projections have been proposed in order to define more robust climate projections that are more representative than those based on a single model [32,48].

3.4. Hydrological Impacts of Climate Change on Groundwater Levels

A simple approach proposed by Scott [49] was applied to assess future CC impacts on global lumped drawdowns due to the future potential rainfall recharge and pumping. Following this approach, simple balance equations were applied in order to assess the global lumped change in hydraulic head (Δh_t) from aquifer storage (ΔS_t), which was calculated as follows.

$$\Delta S_t = R_t - E_t \tag{1}$$

where R_t is the aquifer rainfall recharge and E_t is the aquifer extraction, which can be obtained from Equation (2).

$$E_t = Eag_t - Enonag_t \tag{2}$$

where Eag_t represents the agricultural extractions and $Enonag_t$ the non-agricultural extractions. They were calculated using Equations (3) and (4), respectively.

$$Eag_t = Eag_{t-1}\left[1 + \frac{ET_t - ET_{t-1}}{ET_{t-1}}\right] \tag{3}$$

$$Enonag_t = Enonag_{t-1}\left[1 + \frac{Pop_t - Pop_{t-1}}{Pop_{t-1}}\right] \tag{4}$$

where ET is the evapotranspiration calculated using the Blaney–Criddle method ($ET = p(0.46\mathrm{Tmean} + 8)$) on the basis of monthly temperature in °C (Tmean) and latitude-derived sunshine-hour fraction (p), and Pop is the population of the area.

Finally, the hydraulic head (Δh_t) was calculated using Equation (5).

$$\Delta h_t = \frac{\Delta S_t}{A \times S_y} \tag{5}$$

where A is the aquifer area and S_y the specific yield.

The initial conditions used to simulate the recharge (R_{t-1}) and pumping (agricultural or non-agricultural, Eag_{t-1} and $Enonag_{t-1}$) evolutions were taken from the information included in the Guadalquivir River Basin Plan (2015–2021).

The lumped approach proposed by Scott [49] was also applied to estimate the lumped hydraulic head drawdowns in the reference historical period (1986–2015). The method allowed estimation of the delta change (percentage increase) in the lumped aquifer drawdowns, taking into account the relative difference between the maximum lumped drawdowns in the historical and the future periods (2016–2045). These results were obtained under the assumption that a business-as-usual management scenario will be maintained in the future. The future potential hydraulic head in each piezometer was obtained by applying a delta change correction, using the lumped change to modify the historical evolution (for the reference period) of this variable within the piezometer.

3.5. Propagation of Hydrological Impacts to Subsidence

Simple linear regression models have been defined in order to approximate subsidence as a function of hydraulic head drawdowns in the selected head observation wells. We tested different transformations (Tr(X)) of the explanatory and target variables (logarithm, inverse, square, and square-root mathematical transformations) in order to identify the one that provided the best approximation to the empirical data for this problem (see Table 3).

The models used assume that there is a linear relation between both variables, the dependent variable and the explanatory variable and its transformations, which is reasonable if the deformation is elastic. An analysis of the linear correlation depending on the percentage of clay and silt content in the ground was also proposed in order to identify and discuss when linear regression might represent a better approach.

Table 3. Regression models and transformation of variables applied. The symbol * represents the tested combinations of models and transformation of variables.

			Tr(X)		
Model	-	X^2	**sqrt(X)**	**log(X)**	**1/X**
$Y = a \times X + b$	*	-	-	-	-
$Y = a \times Tr(X) + b$	-	*	*	*	*
$Tr(Y) = a \times Tr(X) + b$	-	*	*	*	*

4. Results and Discussion

For the case study, the future equifeasible ensemble series obtained by applying the bias correction approach showed a mean global temperature increase of 7.72% and a mean global reduction of precipitation of 6.24%. Figure 3 shows the mean historical and future yearly series of precipitation and temperature.

The hydrological approach described in Section 3.4 was employed to propagate the generated future local climatic series to assess future hydrological impacts in terms of future recharge and pumping (Figure 4).

The mean reduction of the recharge and increase of pumping expected for the potential future horizon contemplated were 1.4 Hm^3 year^{-1} and 1 Hm^3 year^{-1}, respectively. In the global budget, the impacts of CC on the recharge will have a higher influence on the future hydraulic head drawdowns.

Taking into account the projected changes in the future recharge and withdrawals within the aquifer, the global lumped hydraulic head drawdowns were obtained by applying the Scott [49] approach. The maximum lumped drawdown obtained in the future was 3.3% greater than the one obtained in the historical period. This relative change was employed to apply a delta change to correct the historical drawdowns in the selected head-observation wells (see Figure 1), obtaining for the future period the values presented in Figure 9.

In order to propagate the impacts of these hydraulic head drawdowns on the subsidence, different regression approaches were tested (see Figure 2). In each piezometer, the coefficient of determination of the calibrated models depended on the selected transformation (see results for the tested models in Figure 5). The two best combinations of model and transformation were $S = a \times P + b$ and $S^2 = a \times P^2 + b$, where S represents the subsidence and P the hydraulic head (Models A and F in Figure 5). The mean R^2 values of these models for the five considered piezometers were 0.60 and 0.73, respectively. These models were employed to predict the impacts of the potential future CC scenario on subsidence. An example of the fit of the regression models for the target variable (subsidence) and the explanatory variable (hydraulic head) is included in Figure 6 for Piezometer 3 and the selected models.

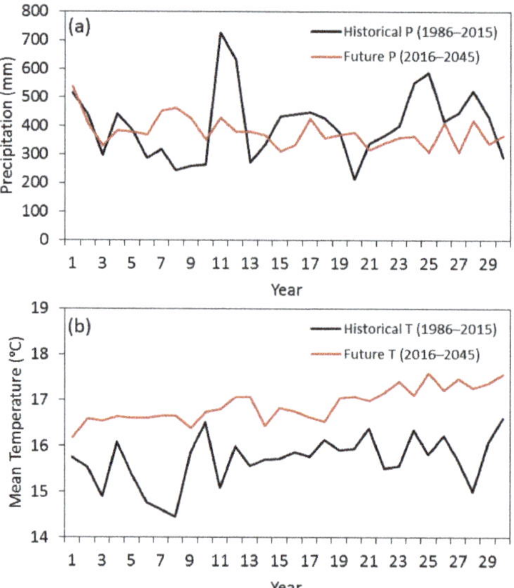

Figure 3. Yearly mean historical and generated future series of precipitation (**a**) and mean temperature (**b**) for the periods 1986–2015 and 2016–2045. The historical data were obtained from the Spain02 dataset and the future series was generated by applying the proposed methodology.

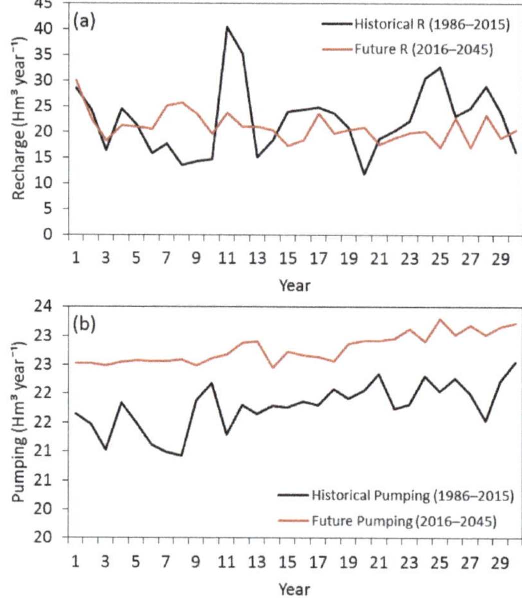

Figure 4. Yearly mean historical and future series of recharge (**a**) and withdrawals (**b**) for the mean year in the periods 1986–2015 and 2016–2045. The historical data were obtained using the methodology proposed by Scott [49] and historical recharge and pumping rates from the Guadalquivir River Basin Plan (2015–2021). The future series were generated by applying the proposed methodology.

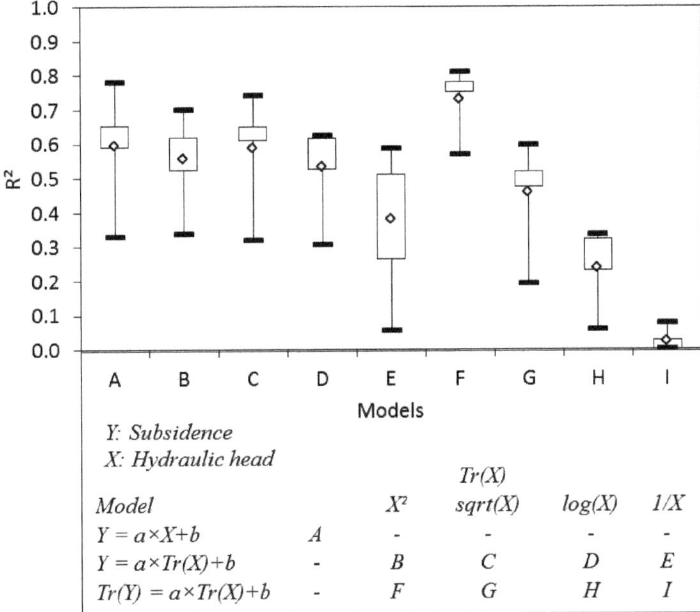

Figure 5. Coefficient of determination (R^2) of the tested models for the five considered piezometers.

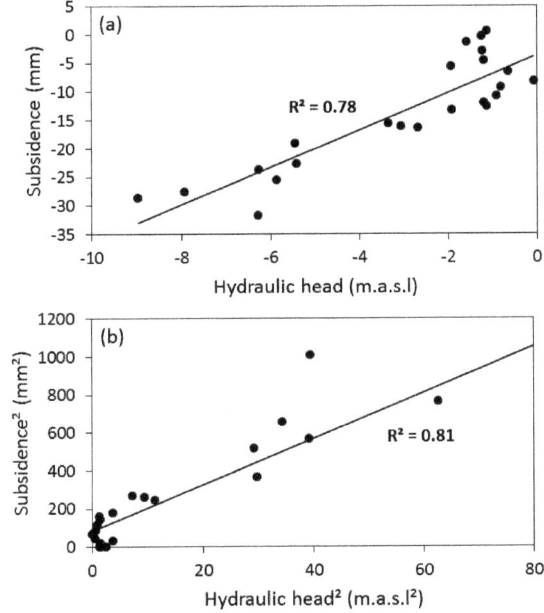

Figure 6. Relationship between the target variable and the explanatory variable for the two best linear regression models ($S = a \times P + b$; $S^2 = a \times P^2 + b$) for Piezometer 3.

The selected models were compared in terms of mean error and mean squared error too (see Figure 7). The two models showed a mean error of 0.0 mm; the model S = a × P + b showed a mean squared error of 20.63 mm^2 and the model S^2 = a × P^2 + b showed a mean squared error of 23.44 mm^2.

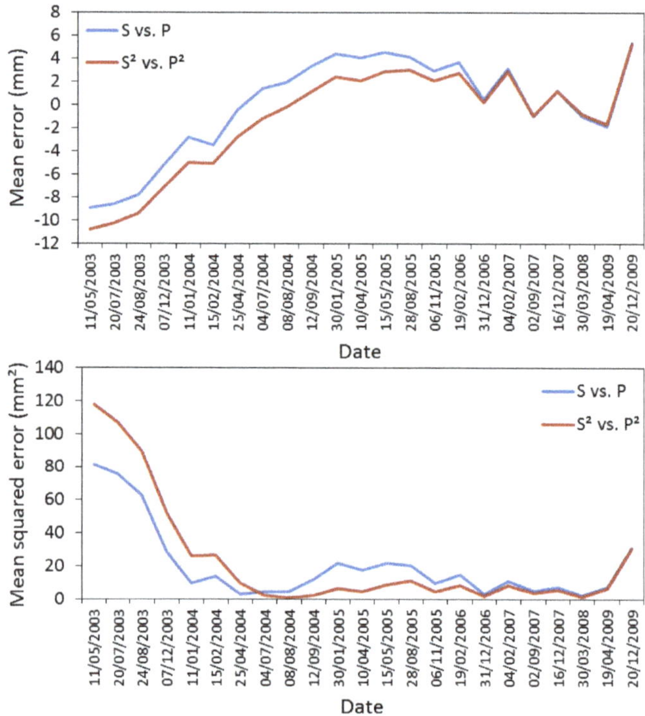

Figure 7. Mean values of the mean error and mean squared error for the five considered piezometers, obtained for the two best linear regression models (S = a × P + b; S^2 = a × P^2 + b).

The granulometry of subsurface sediments has a significant impact on the ground subsidence. We analyzed the influence of the percentage of clay and silt (in the surrounding area of the piezometer) on the historical subsidence rate and the linear behavior of the subsidence, assessed in terms of the R^2 of the regression models.

Figure 8a shows the relationship between the percentage of fine-grained material and the historical subsidence rate, and Figure 8b the coefficient of determination obtained for the best approach for each piezometer vs. the percentage of clay and silt. In general, a higher percentage of fine-grained material was related to a lower subsidence rate, but the correlation of this relationship was poor (R^2 = 0.24). On the other hand, higher coefficients of determination were related with a more linear behavior, which was observed for the cases with lower percentage of clay and silt. Note that in general, the relationship between the percentage of fine-grained material and the coefficients of determination of the linear models had a good correlation (R^2 higher than 0.8). In fact, PSInSAR results showed an inelastic deformation in the aquifer where a higher clay–silt content was identified [15]. Percentage of fine-grained material, thickness, and distribution of lenses significantly affected spatiotemporal subsidence patterns, which was in agreement with the results observed by other authors [10,50].

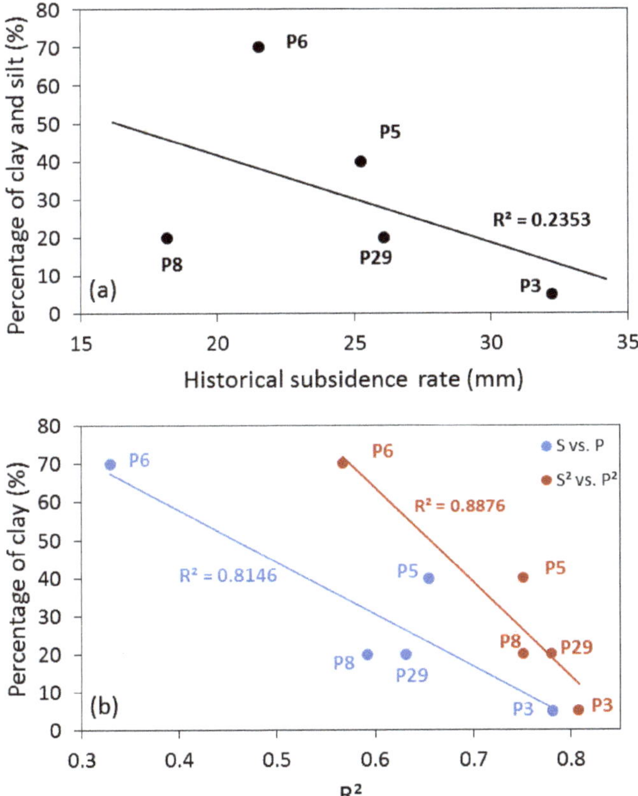

Figure 8. Relationship between the percentage of clay and silt and the historical subsidence rate (**a**) and the correlation coefficients obtained for the two best linear regression models (S = a × P + b; $S^2 = a \times P^2 + b$) (**b**).

For Piezometer 6, which was the one with highest percentage of clay (70%), we obtained very low coefficient of determination (0.33 and 0.57 for the models S = a × P + b and $S^2 = a \times P^2 + b$, respectively) and the linear regression approach was not appropriate. For this reason, we have not included the results for the assessment of the future subsidence for this observation well.

The propagation of the impacts of the potential future CC scenario on subsidence (Figure 9) showed important increases of the maximum subsidence (55.3% and 52.7% for the models S = a × P + b and $S^2 = a \times P^2 + b$, respectively) with respect to the historical maximum observed values. The highest increase of the maximum subsidence (68.3% and 65.7% for the models S = a × P + b and $S^2 = a \times P^2 + b$, respectively) with respect to the historical maximum observed values occurred in Piezometer 5, which was located in the western area where the percentage of clay was 40%.

In terms of mean subsidence (see Figure 10) the mean increase of subsidence was 4.1 mm and 4.5 mm for the models S = a × P + b and $S^2 = a \times P^2 + b$, respectively. The piezometer with the highest increase of mean subsidence was Piezometer 3, with 5.4 mm and 5.9 mm for the models S = a × P + b and $S^2 = a \times P^2 + b$, respectively. This piezometer showed the highest variability of subsidence in the historical and future periods, and had the lowest content of clay and silt.

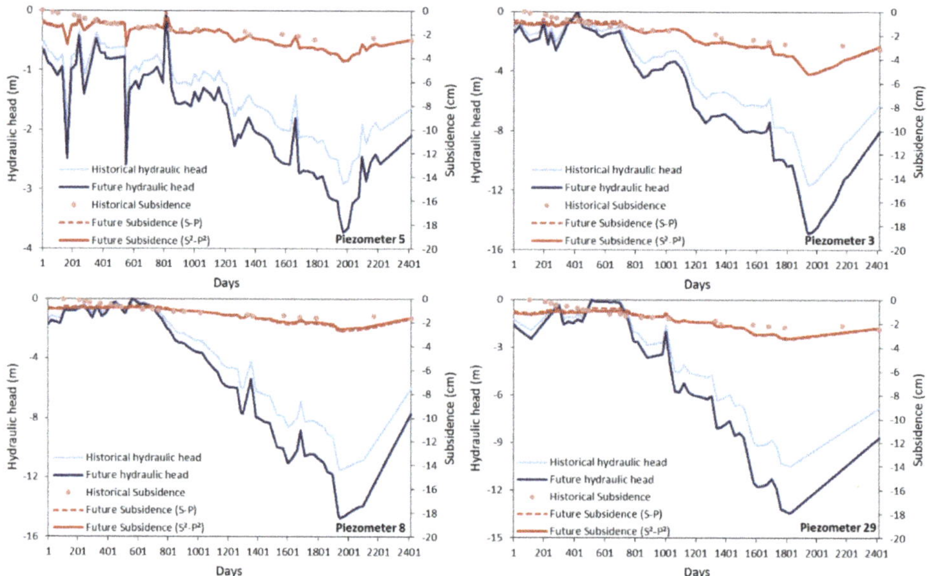

Figure 9. Historical and future subsidence and hydraulic head for the two best linear regression models ($S = a \times P + b$; $S^2 = a \times P^2 + b$).

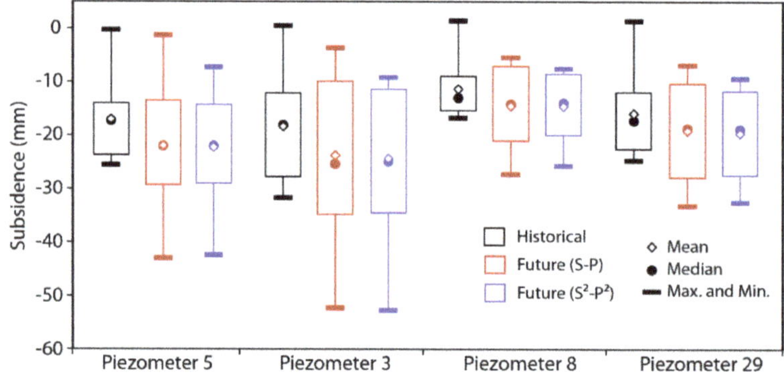

Figure 10. Historical and future subsidence for the two best linear regression models ($S = a \times P + b$; $S^2 = a \times P^2 + b$).

Hypotheses Assumed and Limitations

The assessment performed assumed some hypotheses or made simplifications as follows.

Climate change scenarios

-   We generated future local climatic scenarios only for the horizon 2016–2045, assuming the most pessimistic emissions scenario (RCP 8.5) and applying a simple statistical correction (first- and second-moment correction) to correct the observed biases.
-   An equifeasible ensemble of the potential future local climate scenarios was proposed to define a more representative scenarios by combining projections of different climatic models.

Propagation of climate change in terms of hydrological impacts

- A simple approach proposed by Scott [49] was applied in order to perform the assessment of future CC impacts on global lumped drawdowns. It has the advantages that it does not require the use of a previously calibrated distributed model, and can be applied in cases with limited information. A delta change approach defined from this lumped variable was employed to assess drawdowns in the piezometers by correcting the historical series. More precise results could be obtained in cases where a physically distributed model is directly used to propagate climate change impacts.
- This work focused exclusively on the impacts produced by the reduction in rainfall recharge and the increase of pumping due to potential future local climate scenarios and changes in population.
- We assumed a business-as-usual management scenario to assess the impacts of future potential local climate scenarios on subsidence. Other management scenarios could be considered in order to assess the benefit of potential adaptation strategies.

Propagation of hydrological impacts to subsidence

- We assumed a linear relationship between the hydraulic head drawdowns and the subsidence in the piezometers. This allowed us to use a simple regression model. We also assumed that the model was valid in the range in which the future assessment was performed.

5. Conclusions

A method to assess impact of potential future CC scenarios on land subsidence related to groundwater level depletion in detrital aquifers was described and applied in a case study. It does not require a previous distributed groundwater flow model of the aquifer, which is an important advantage for its applicability in cases with limited information. Taking into account the uncertainties around future potential CC scenarios, it could provide a useful first assessment of its impacts on subsidence. It allows analyses of wide areas to be performed in order to identify potential hot spots that require more exhaustive analysis. It will help to assess sustainable adaptation strategies in identified vulnerable areas, taking into account subsidence issues during drought-critical periods. The methodology was applied in the Vega de Granada aquifer (Granada, SE Spain). Good correlation between groundwater level depletion and the subsidence was obtained in the wells where the percentage of clay was below 50%. The analysis of results showed that, assuming a business-as-usual management scenario, the impacts of CC on subsidence would be very significant for the case study. The mean increase of the maximum subsidence rates in the considered wells for the future horizon (2016–2045) and the RCP scenario 8.5 was 54%. In order to avoid undesirable consequences/risks as observed in other regions worldwide (Jakarta, Ho Chi Min City, and Bangkok, among others) where land subsidence is causing severe impacts—permanent inundation of land, aggravated flooding, changes in topographic gradients, rupture of the land surface, structural damage to buildings and infrastructures, and reduced capacity of aquifers to store water—some adaptation strategies should be applied to control and minimize the land subsidence caused by groundwater withdrawals.

Author Contributions: D.P.-V. and R.M.M. conceived and designed the research; P.E. analyzed the subsidence data; A.-J.C.-L. analyzed the data and conducted the experiments. All authors contributed to writing the manuscript. All authors have read and agreed to the published version of the manuscript.

Funding: This research was partially supported by the research projects SIGLO-AN (RTI2018-101397-B-I00) from the Spanish Ministry of Science, Innovation and Universities (Programa Estatal de I+D+I orientada a los Retos de la Sociedad) and the GeoE.171.008-TACTIC project from GeoERA organization funded by European Union's Horizon 2020 research and innovation program.

Acknowledgments: We would like to thank the Spain02 and CORDEX projects, the ENVISAT, Cosmo-SkyMed and Sentinel satellites, the Spanish Statistical Office, the Spanish official network for monitoring the quantitative state of groundwater and the Guadalquivir River Basin authority for the data provided for this study.

Conflicts of Interest: The authors declare no conflict of interest.

References

1. Abidin, H.Z.; Andreas, H.; Gumilar, I.; Fukuda, Y.; Pohan, Y.; Deguchi, T. Land subsidence of Jakarta (Indonesia) and its relation with urban development. *Nat. Hazards* **2011**, *59*, 1753. [CrossRef]
2. Lixin, Y.; Jie, W.; Chuanqing, S.; Guo, J.W.; Yanxiang, J.; Liu, B. Land Subsidence Disaster Survey and Its Economic Loss Assessment in Tianjin, China. *Nat. Hazards Rev.* **2010**, *11*, 35–41. [CrossRef]
3. Pacheco-Martínez, J.; Hernández-Marín, M.; Burbey, T.J.; González-Cervantes, N.; Ortiz, J.; Solís-Pinto, A. Land subsidence and ground failure associated to groundwater exploitation in the Aguascalientes Valley, México. *Eng. Geol.* **2013**, *164*, 172–186. [CrossRef]
4. Wang, G.Y.; Zhang, D.; Feng, J.S.; Chen, M.Z.; Shan, W.H. Land subsidence due to deep groundwater withdrawal in northern Yangtze river delta area. In *Engineering Geology for Society and Territory, Urban Geology, Sustainable Planning and Landscape Exploitation*; Springer: Cham, Switzerland, 2015; Volume 5. [CrossRef]
5. Ye, S.; Xue, Y.; Wu, J.; Yan, X.; Yu, J. Progression and mitigation of land subsidence in China. *Hydrogeol. J.* **2016**, *24*, 685–693. [CrossRef]
6. Faunt, C.C.; Sneed, M.; Traum, J.; Brandt, J.T. Water availability and land subsidence in the Central Valley, California, USA. *Hydrogeol. J.* **2016**, *24*, 675–684. [CrossRef]
7. De Luna, R.M.R.; Garnés, S.J.A.; Cabral, J.J.S.P.; Melo dos Santos, S. Groundwater overexploitation and soil subsidence monitoring in Recife plain (Brazil). *Nat. Hazards* **2017**, *86*, 1363. [CrossRef]
8. Erkens, G.; Bucx, T.; Dam, R.; de Lange, G.; Lambert, J. Sinking coastal cities. In *Prevention and Mitigation of Natural and Anthropogenic Hazards Due to Land Subsidence, Proceedings of the IAHS, Nagoya, Japan, 12 November 2015*; Daito, K., Galloway, D., Eds.; Copernicus Publications: Göttingen, Germany, 2015; Volume 372, pp. 189–198.
9. Galloway, D.L.; Burbey, T.J. Review: Regional land subsidence accompanying groundwater extraction. *Hydrogeol. J.* **2011**, *19*, 1459–1486. [CrossRef]
10. Galloway, D.L.; Erkens, G.; Kuniansky, E.; Rowland, J. Preface: Land subsidence processes. *Hydrogeol. J.* **2016**, *24*, 547–550. [CrossRef]
11. Tomás, R.; Romero, R.; Mulas, J.; Marturiá, J.; Mallorqui, J.; López-Sánchez, J.M.; Hererra, G.; Gutiérrez, F.; González, P.; Fernández, J.; et al. Radar interferometry techniques for the study of ground subsidence phenomena: A review of practical issues through cases in Spain. *Environ. Earth Sci.* **2014**, *71*, 163–181. [CrossRef]
12. Chaussard, E.; Wdowinski, S.; Cabral-Cano, E.; Amelung, F. Land subsidence in central Mexico detected by ALOS InSAR time-series. *Remote Sens. Environ.* **2014**, *140*, 94–106. [CrossRef]
13. Mahmud, M.U.; Yakubu, T.A.; Oluwafemi, O.; Sousa, J.J.; Ruiz-Armenteros, A.M.; Arroyo-Parras, J.G.; Bakoň, M.; Lazecky, M.; Perissin, D. Application of Multi-Temporal Interferometric Synthetic Aperture Radar (MT-InSAR) technique to Land Deformation Monitoring in Warri Metropolis, Delta State, Nigeria. *Procedia Comput. Sci.* **2016**, *100*, 1220–1227. [CrossRef]
14. Chen, M.; Tomás, R.; Li, Z.; Motagh, M.; Li, T.; Hu, L.; Gong, H.; Li, X.; Yu, J.; Gong, X. Imaging Land Subsidence Induced by Groundwater Extraction in Beijing (China) Using Satellite Radar Interferometry. *Remote Sens.* **2016**, *8*, 468. [CrossRef]
15. Mateos, R.M.; Ezquerro, P.; Luque, J.A.; Béjar-Pizarro, M.; Notti, D.; Azañón, J.M.; Montserrat, O.; Herrera, G.; Fernández-Chacón, F.; Peinado, T.; et al. Multiband PSInSAR and long-period monitoring of land subsidence in a strategic detrital aquifer (Vega de Granada, SE Spain): An approach to support management decisions. *J. Hydrol.* **2017**, *533*, 71–87. [CrossRef]
16. Meisina, C.; Zucca, F.; Notti, D.; Colombo, A.; Cicchi, A.; Savio, G.; Giannico, C.; Bianchi, M. Geological interpretation of PSInSAR Data at regional scale. *Sensors* **2008**, *8*, 7469–7492. [CrossRef] [PubMed]
17. Burbey, T.J.; Zhang, M. Inverse modeling using PSInSAR for improved calibration of hydraulic parameters and prediction of future subsidence for Las Vegas Valley, USA. In Proceedings of the International Association of Hydrological Sciences (IAHS '15), Nagoya, Japan, 12 November 2015; Volume 372, pp. 411–416.
18. Leake, S.A.; Galloway, D.L. *MODFLOW Ground-Water Model—User Guide to the Subsidence and Aquifer-System Compaction Package (SUB-WT) for Water-Table Aquifers*; Techniques and Methods; U.S. Geological Survey: Reston, VA, USA, 2007; p. 42.

19. Rahmati, O.; Falah, F.; Naghibi, S.; Biggs, T.; Soltani, M.; Deo, R.; Cerdà, A.; Mohammadi, F.; Tien Bui, D. Land subsidence modelling using tree-based machine learning algorithms. *Sci. Total Environ.* **2019**, *672*, 239–252. [CrossRef] [PubMed]

20. Li, J. A nonlinear elastic solution for 1-D subsidence due to aquifer storage and recovery applications. *Hydrogeol. J.* **2003**, *11*, 646–658. [CrossRef]

21. Ezquerro, P.; Guardiola-Albert, C.; Herrera, G.; Fernandez-Merodo, J.A.; Bejar, M.; Bonì, R. Groundwater and Subsidence Modeling Combining Geological and Multi-Satellite SAR Data over the Alto Guadalentín Aquifer (SE Spain). *Geofluids* **2017**, 1–17. [CrossRef]

22. IPCC. *Climate Change 2014: Impacts, Adaptation, and Vulnerability. Part A: Global and Sectoral Aspects*; Field, C.B., Barros, V.R., Dokken, D.J., Mach, K.J., Mastrandrea, M.D., Bilir, T.E., Chatterjee, M., Ebi, K.L., Estrada, Y.O., Genova, R.C., et al., Eds.; Contribution of working group II to the fifth assessment report of the intergovernmental panel on climate change; Cambridge University Press: Cambridge, UK; New York, NY, USA, 2014.

23. Dragoni, W.; Sukhija, B.S. Climate change and groundwater—A short review. In *Climate Change and Groundwater*; Dragoni, W., Sikhija, B.S., Eds.; Special Publications; Geological Society: London, UK, 2008; Volume 288, pp. 1–12.

24. Collados-Lara, A.J.; Pulido-Velazquez, D.; Pardo-Igúzquiza, E. An Integrated Statistical Method to Generate Potential Future Climate Scenarios to Analyse Droughts. *Water* **2018**, *10*, 1224. [CrossRef]

25. UN. *Guidance on Water y Adaptation to Climate Change*; ECE/MP.WAT/30; United Nations: Geneva, Switzerland, 2009.

26. Teng, J.; Chiew, F.; Vaze, J.; Marvanek, S.; Kirono, D. Estimation of Climate Change Impact on Mean Annual Runoff across Continental Australia Using Budyko and Fu Equations and Hydrological Models. *J. Hydrometeorol.* **2012**, *13*, 1094–1106. [CrossRef]

27. Verzano, K.; Bärlund, I.; Flörke, M.; Lehner, B.; Kynast, E.; Voß, F.; Alcamo, J. Modeling variable river flow velocity on continental scale: Current situation and climate change impacts in Europe. *J. Hydrol.* **2012**, *424*, 238–251. [CrossRef]

28. Pulido-Velazquez, D.; Collados-Lara, A.J.; Alcalá, F.J. Assessing impacts of future potential climate change scenarios on aquifer recharge in continental Spain. *J. Hydrol.* **2018**, *567*, 803–819. [CrossRef]

29. Barranco, L.M.; Álvarez-Rodríguez, J.; Olivera, F.; Potenciano, A.; Quintas, L.; Estrada, F. Assessment of the Expected Runoff Change in Spain Using Climate Simulations. *J. Hydrol. Eng.* **2014**, *19*, 1481–1490. [CrossRef]

30. Escriva-Bou, A.; Pulido-Velazquez, M.; Pulido-Velazquez, D. The Economic Value of Adaptive Strategies to Global Change for Water Management in Spain's Jucar Basin. *J. Water Resour. Plan. Manag.* **2017**, *143*, 5. [CrossRef]

31. Pulido-Velazquez, D.; Garrote, L.; Andreu, J.; Martin-Carrasco, F.J.; Iglesias, A. A methodology to diagnose the effect of climate change and to identify adaptive strategies to reduce its impacts in conjunctive-use systems at basin scale. *J. Hydrol.* **2011**, *405*, 110–122. [CrossRef]

32. Pulido-Velázquez, D.; García-Aróstegui, J.L.; Molina, J.L.; Pulido-Velázquez, M. Assessment of future groundwater recharge in semi-arid regions under climate change scenarios (Serral-Salinas aquifer, SE Spain). Could increased rainfall variability increase the recharge rate? *Hydrol. Process.* **2015**, *29*, 828–844. [CrossRef]

33. Hanson, R.; Flint, A.; Flint, L.; Faunt, C.; Schmid, W.; Dettinger, M.; Leake, S.; Cayan, R.D. Integrated simulation of consumptive use and land subsidence in the Central Valley, California, for the past and for a future subject to urbanization and climate change. In Proceedings of the Eighth International Symposium on Land Subsidence (EISOLS), Queretaro, Mexico, 17–22 October 2010; IAHS Publ.: Wallingford, UK, 2010; Volume 339, pp. 17–22.

34. Brouns, K.; Eikelboom, T.; Jansen, P.C.; Janssen, R.; Kwakernaak, C.; van den Akker, J.J.H.; Verhoeven, J.T.A. Spatial Analysis of Soil Subsidence in Peat Meadow Areas in Friesland in Relation to Land and Water Management, Climate Change, and Adaptation. *Environ. Manag.* **2014**, 55. [CrossRef]

35. Zhou, Y.; Li, W. A review of regional groundwater flow modeling. *Geosci. Front.* **2011**, *2*, 205–214. [CrossRef]

36. Skaugen, T.; Peerebom, I.O.; Nilsson, A. Use of a parsimonious rainfallrunoff model for predicting hydrological response in ungauged basins. *Hydrol. Process.* **2015**, *29*, 1999–2013. [CrossRef]

37. Tornqvist, T.E.; Wallace, D.J.; Storms, J.E.A.; Wallinga, J.; van Dam, R.L.; Blaauw, M.; Derksen, M.S.; Klerks, C.J.W.; Meijneken, C.; Snijders, E.M.A. Mississippi delta subsidence primarily caused by compaction of Holocene strata. *Nat. Geosci.* **2008**, *1*, 173–176. [CrossRef]

38. Parcharidis, I.; Kourkouli, P.; Karymbalis, E.; Foumelis, M.; Karathanassi, V. Time Series Synthetic Aperture Radar Interferometry for Ground Deformation Monitoring over a Small Scale Tectonically Active Deltaic Environment (Mornos, Central Greece). *J. Coast. Res.* **2013**, *29*, 325–338. [CrossRef]

39. Pulido-Velazquez, D.; Sahuquillo, A.; Andreu, J.; Pulido-Velazquez, M. A general methodology to simulate groundwater flow of unconfined aquifers with a reduced computational cost. *J. Hydrol.* **2007**, *338*, 42–56. [CrossRef]

40. Llopis-Albert, C.; Pulido-Velazquez, D. Using MODFLOW code to approach transient hydraulic head with a sharp-interface solution. *Hydrol. Process.* **2015**, *29*, 2052–2064. [CrossRef]

41. Kohfahl, C.; Sprenger, C.; Herrera, J.B.; Meyer, H.; Chacón, F.F.; Pekdeger, A. Recharge sources and hydrogeochemical evolution of groundwater in semiarid and karstic environments: A field study in the Granada Basin (Southern Spain). *Appl. Geochem.* **2008**, *23*, 846–862. [CrossRef]

42. Azañón, J.M.; Azor, A.; Booth-Rea, G.; Torcal, F. Small-scale faulting, topographic steps and seismic ruptures in the Alhambra (Granada, southeast Spain). *J. Quat. Sci.* **2004**, *19*, 219–227. [CrossRef]

43. Luque-Espinar, J.A.; Chica-Olmo, M.; Pardo-Igúzquiza, E.; García-Soldado, M.J. Influence of climatological cycles on hydraulic heads across a Spanish aquifer. *J. Hydrol.* **2008**, *354*, 33–52. [CrossRef]

44. Herrera, S.; Fernández, J.; Gutiérrez, J.M. Update of the Spain02 Gridded Observational Dataset for Euro-CORDEX evaluation: Assessing the Effect of the Interpolation Methodology. *Int. J. Climatol.* **2016**, *36*, 900–908. [CrossRef]

45. CORDEX PROJECT. The Coordinated Regional Climate Downscaling Experiment (CORDEX). Program Sponsored by World Climate Research Program (WCRP). 2013. Available online: https://www.wcrp-climate.org/news/wcrp-newsletter/wcrp-news-articles/1347-wcrp-spotlight-the-coordinated-regional-climate-downscaling-experiment-cordex (accessed on 11 January 2020).

46. Devanthéry, N.; Crosetto, M.; Monserrat, O.; Cuevas-González, M.; Crippa, B. An approach to persistent scatterer interferometry. *Remote Sens.* **2014**, *6*, 6662–6679. [CrossRef]

47. Barra, A.; Monserrat, O.; Mazzanti, P.; Esposito, C.; Crosetto, M.; Scarascia Mugnozza, G. First insights on the potential of Sentinel-1 for landslides detection. *Geomat. Nat. Hazards Risk* **2016**, *7*, 1874–1883. [CrossRef]

48. Hashmi, M.Z.; Shamseldin, A.Y.; Melville, B.W. Statistically downscaled probabilistic multi-model ensemble projections of precipitation change in a watershed. *Hydrol. Process.* **2013**, *27*, 1021–1032. [CrossRef]

49. Scott, C.A. The Water-Energy-Climate Nexus: Resources and Policy Outlook for Aquifers in Mexico. *Water Resour. Res.* **2011**, *47*, 1–18. [CrossRef]

50. Miller, M.M.; Shirzaei, M.; Argus, D. Aquifer Mechanical Properties and Decelerated Compaction in Tucson, Arizona. *J. Geophys. Res. Solid Earth* **2017**, *122*, 8402–8416. [CrossRef]

Article

Numerical Approaches for Estimating Daily River Leakage from Arid Ephemeral Streams

Leilei Min [1,2,†], Peter Yu. Vasilevskiy [3,†], Ping Wang [2,4,*], Sergey P. Pozdniakov [3] and Jingjie Yu [2,4]

1   Hebei Key Laboratory of Water-Saving Agriculture, Key Laboratory of Agricultural Water Resources, The Innovative Academy of Seed Design, Center for Agricultural Resources Research, Chinese Academy of Sciences, Shijiazhuang 050021, China; llmin@sjziam.ac.cn
2   Key Laboratory of Water Cycle and Related Land Surface Processes, Institute of Geographic Sciences and Natural Resources Research, Chinese Academy of Sciences, 11A, Datun Road, Chaoyang District, Beijing 100101, China; yujj@igsnrr.ac.cn
3   Department of Hydrogeology, Lomonosov Moscow State University, GSP-1, Leninskie Gory, Moscow 119899, Russia; valenciacf@mail.ru (P.Y.V.); sppozd@geol.msu.ru (S.P.P.)
4   College of Resources and Environment, University of the Chinese Academy of Sciences, Beijing 100049, China
*   Correspondence: wangping@igsnrr.ac.cn; Tel.: +86-10-6488-1192
†   These authors contributed equally to this work.

Received: 27 December 2019; Accepted: 9 February 2020; Published: 12 February 2020

Abstract: Despite the significance of river leakage to riparian ecosystems in arid/semi-arid regions, a true understanding and the accurate quantification of the leakage processes of ephemeral rivers in these regions remain elusive. In this study, the patterns of river infiltration and the associated controlling factors in an approximately 150-km section of the Donghe River (lower Heihe River, China) were revealed using a combination of field investigations and modelling techniques. The results showed that from 21 April 2010 to 7 September 2012, river water leakage accounted for 33% of the total river runoff in the simulated segments. A sensitivity analysis showed that the simulated infiltration rates were most sensitive to the aquifer hydraulic conductivity and the maximum evapotranspiration (ET) rate. However, the river leakage rate, i.e., the ratio of the leakage volume to the total runoff volume, of a single runoff event relies heavily on the total runoff volume and river flow rate. In addition to the hydraulic parameters of riverbeds, the characteristics of ET parameters are equally important for quantifying the flux exchange between arid ephemeral streams and underlying aquifers. Coupled surface/groundwater models, which aim to estimate river leakage, should consider riparian zones because these areas play a dominant role in the formation of leakage from the river for recharging via ET. The results of this paper can be used as a reference for water resource planning and management in regulated river basins to help maintain riparian ecosystems in arid regions.

Keywords: river-aquifer interaction; numerical simulation; sensitivity analysis; MODFLOW; Heihe River

---

1. Introduction

Surface water and groundwater are important components of the terrestrial water cycle, and their interaction forms the surface morphology, controls the material and energy fluxes in the subsurface zone, and affects the riparian ecosystem [1,2]. However, as indicated by Sophocleous [3], the interactions between surface water and groundwater are complex, and obtaining a deep understanding of these interactions in relation to the climate, landform features, geology, and biotic factors remains a great challenge. Clearly, the exchange between surface water and groundwater is likely to become even more challenging due to the impacts of human activities and climate change [4,5], which caused the disappearance of approximately 90,000 km^2 of permanent surface water between 1984 and 2015 [6].

Given that a streambed acts as the physical interface between the surface and subsurface of a stream [7,8], the hydraulic properties of streambeds mainly control the interactions between the stream and the underlying aquifer [4,9–11]. However, as confirmed by numerous field investigations (e.g., [12–14]), the hydraulic properties of streambeds usually exhibit large spatial and temporal variations, which are mainly caused by continuous changes in the streambed properties (e.g., the topography or hydraulic conductivity) during erosion and sedimentation processes [7]. Significant changes in the hydraulic properties of streambeds may even occur during short flooding events [15]. Additionally, the thermal dynamics of streambeds induced by diurnal and seasonal fluctuations in the stream water temperature also greatly influence the hydraulic properties of the streambeds [16–18]. For intermittent rivers, which constitute more than 30% of the total length and discharge of the global river network [2,19], the hydraulic properties of streambeds are even more variable due to constant alternation between dry and wet conditions [1,16].

The states of the connections between streams and underlying aquifers exert another important influence on the flux exchange between surface water and groundwater [4]. For losing streams, when stream-aquifer systems are transformed from connected to disconnected systems, the lateral flow induced by capillarity or heterogeneity plays a vital role in the stream water and groundwater interaction [20–22]. Brunner et al. [23] provided a theoretical criterion for justifying the connection/disconnection states between a stream and the underlying aquifer and suggested that the disconnection problem could be solved via a fully coupled, variably saturated flow model. Therefore, in addition to the broad range of field methods (e.g., [24–27]) and associated analytical solutions e.g., [17,28,29], numerical simulations have been widely applied to investigate stream–aquifer interactions at different scales because these simulations can analyze the influences of transient flows and streambed heterogeneity on surface–groundwater exchanges [4,30].

Recently, interest in the interactions between intermittent streams and groundwater in arid and semi-arid regions has been continuously growing due to the unique role of these interactions in shaping fragile riparian ecosystems (e.g., [27,31–35]). It is clear that stream water leakage is the dominant recharge mechanism in such regions; however, the infiltration processes during various stream discharge patterns and the factors that control the stream–aquifer interactions in typical losing connected/disconnected river systems remain unclear. In this context, it is critical to quantify the potential impacts of the variations in the stream width and leakage coefficient [36,37], which vary greatly for intermittent rivers in arid regions [1], on stream–aquifer interactions.

The lower Heihe River Basin represents a typical extremely arid region in north-western China, where the mean annual precipitation is less than 50 mm, while the mean annual evaporation can reach 1500 mm [38–40]. The lower Heihe River is characterized by intermittent streams, and the streambeds usually remain dry from April to June [41]. The hydraulic property dynamics of these streambeds were investigated in detail in recent studies (e.g., [13,16,41]), and the monthly river leakage was approximately estimated using the River (RIV) package of MODFLOW-2005 [42,43]. While this estimation was based on regional groundwater modelling, it lacked a detailed analysis of the influence of the stream/streambed dynamics on the water exchange between the stream and aquifer. Therefore, to fill this gap, the objectives of this study were to (1) quantify the daily river leakage rates by numerically simulating the flux exchange between the rivers and the underlying aquifers and (2) identify the predominant factors that control the river–aquifer interactions in intermittent dryland rivers using parameter sensitivity analysis.

2. Materials and Methods

2.1. Study Area

The study area is located in the lower reaches of the second largest inland river in north-western China (Figure 1) and is characterized by a hyper-arid climate with an annual precipitation of only 35 mm and an annual potential evaporation of approximately 1500 mm [38,40]. Over the period of 1961–2015, the mean annual air temperature was +9.09 °C, with a minimum monthly mean air temperature of −11.23 °C in January and a maximum monthly mean air temperature of +27.05 °C in July [16]. The topography of this area gradually declines from the southwest to the northeast, with an average slope of 1–3‰, and the elevation is between 1127 and 820 m [44]. The dominant landscape is the Gobi Desert, which is composed of wind-eroded hilly land, desert, and alkaline soils [45].

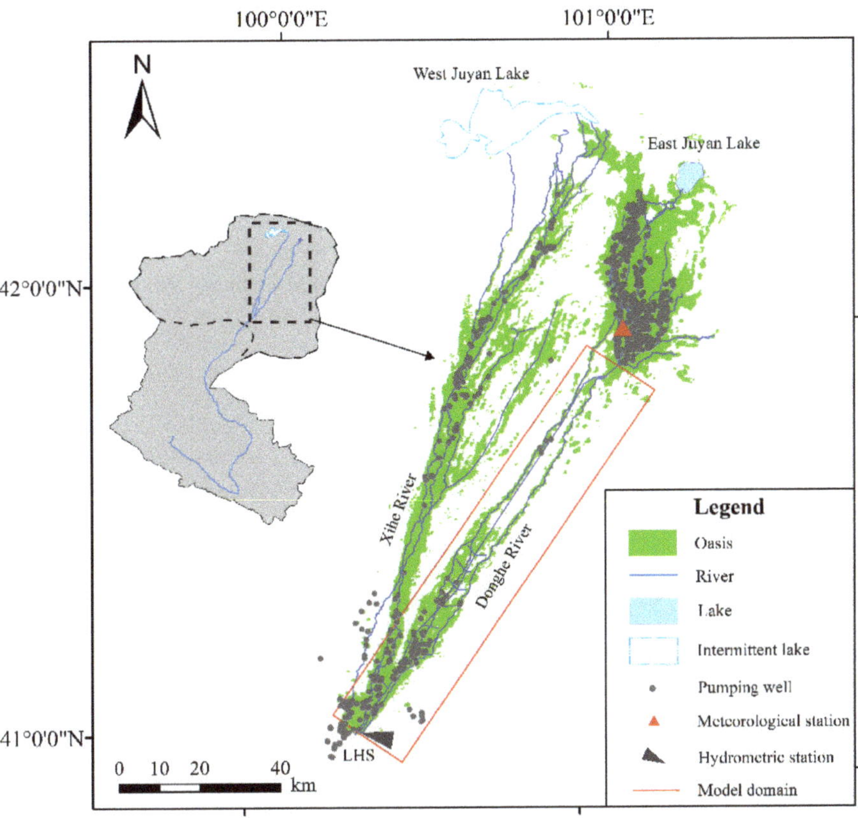

Figure 1. Study area and simulation domain. LHS represents the Langxinshan hydrometric station.

The study area is located in a regional tectonic basin where the bedrock is composed of Sinian (Z) and Late Jurassic (J_3) formations. From the southwest to the northeast of the study area, the regional Quaternary (Q) aquifer system varies from a zone that consists of unconfined gravel and pebbles to a multi-layered zone that consists of sand and silt with a depth of several hundred metres [46,47]. In general, the phreatic aquifer is recharged by river water, and the groundwater flows from the southwest towards the terminal lakes and is then discharged via evaporation [39,48].

The lower Heihe River, which is divided into two losing streams at the Langxinshan hydrometric station (LHS), i.e., the Donghe and Xihe rivers, flows through the Gobi Desert before entering terminal lakes (the East and West Juyan lakes) (Figure 1). At the LHS, the surface flow to the Donghe and Xihe rivers is regulated by a system of sluices. These river systems are the primary sources of shallow groundwater recharging via riverbed infiltration [39,49] due to the relatively high vertical hydraulic conductivity [13]. The limited vegetation in the region is distributed along the rivers and relies on surface water and shallow groundwater in the riparian zone for sustenance [50–52]. More detailed descriptions of the study area are presented in Wang, Yu, Zhang, and Liu [48]; Wang, Yu, Pozdniakov, Grinevsky, and Liu [39]; and Yao, Zheng, Liu, Cao, Xiao, Li, and Li [42].

2.2. Simulation of River Water and Groundwater Interactions

2.2.1. Simulation Domain and Boundary Conditions

As noted by Wang, Pozdniakov and Vasilevskiy [16], approximately 71% of the total runoff was allocated to the Donghe River at the LHS from 1988 to 2015 (Figure 1). Additionally, the characteristics of the streambed sediment formation in the Donghe River are typical of the study area [13]. Therefore, the Donghe River was selected to analyze the water exchange processes between the river and the underlying aquifer. The simulation domain was determined according to the surface water and groundwater interaction zone during river flow events. Based on previous studies (e.g., [39,42,48,53]), the eastern and western boundaries of the model were determined by the regional flow direction, which was generally parallel to the river channel and approximately 10 km from the Donghe River. Thus, the eastern and western boundaries were generalized as no-flow boundaries. The southern and northern boundaries were at the LHS and Angcizha, respectively. The simulation domain is presented in Figure 1, and the total area of the simulation region was 2306.25 km^2.

To focus on the river–aquifer interactions, the upper unconfined aquifer was simulated as spatially heterogeneous single-layer aquifer system. The top and bottom elevations of the aquifer are shown in Figure 2b,c. Based on geochemical data [48,54] and the regional modelling of the groundwater flow system [42], the southern and northern boundaries of the model were designed as general head boundaries (GHBs) [55] in accordance with lateral groundwater flow from adjacent groundwater basins into the study area. The flux at the top boundary was determined according to the meteorological conditions, surface water and irrigation water infiltration, and evapotranspiration (ET) processes. The bottom of the aquifer was considered impermeable; therefore, the bottom boundary was designed as a no-flow boundary.

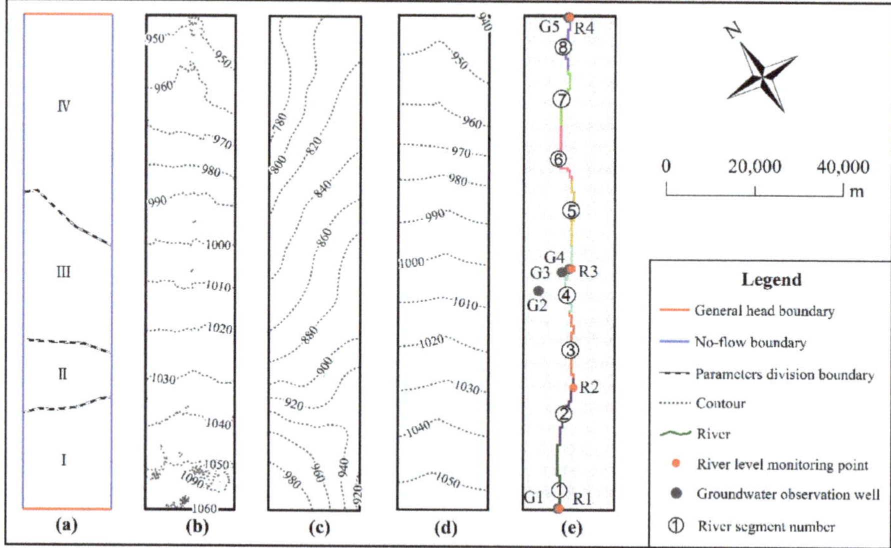

Figure 2. Model setup: (**a**) parameter division; (**b**) land surface elevation; (**c**) bottom elevation of the aquifer; (**d**) initial water levels; and (**e**) river segment division and monitoring sites.

2.2.2. Field Observations and Model Parameterization

The daily groundwater levels were monitored at five observation wells (G1, G2, G3, G4, and G5) located along the riverbank (Figure 2e). The daily river water levels were also observed at wells R1, R2, R3, and R4, which were installed along the Donghe River (Figure 2e). The water levels were recorded by Schlumberger Mini-Diver pressure transducers (Eijkelcamp, EM Giesbeek, The Netherlands). The water level measurements compensated for barometric changes, which were measured by Schlumberger Baro-Divers. The uncertainty of the water level measurements was ±5 mm. In addition, the daily streamflow data at the LHS and daily E-601 pan evaporation rates at the Ejina meteorological station (Figure 1) were available for the period of 2010–2012.

Based on previous hydrogeological investigations [46] and numerical modelling [43], four aquifer zones with different hydraulic parameters were defined (Figure 2a, Table 1). The initial values of hydraulic conductivity varied from 6 to 21 m/day, and the initial specific yield was set to 0.15.

Table 1. Initial and calibrated values of the hydraulic conductivity (K) and specific yield (S_y) of the aquifer.

Zone No.	Dominant Material	Initial Value		Calibrated Value	
		K, m/Day	S_y, -	K, m/Day	S_y, -
1	Coarse sand, gravel	21	0.15	28	0.20
2	Coarse sand	16	0.15	26	0.22
3	Medium sand	11	0.15	23	0.17
4	Fine sand	6	0.15	17	0.15

Based on the riverbed hydraulic conductivities measured in previous studies [13,16], we divided the Donghe River into eight segments, as shown in Figure 2. The parameters of each river segment, including the river width, thickness, and hydraulic conductivity, are listed in Table 2.

Table 2. River width (L_r), riverbed k_0/m_0, and riverbed hydraulic conductance (C) of the river segments.

River Segment No.	Segment Length, m	L_r, m	k_0/m_0, Day^{-1}	C, m^2/Day	
				Initial	Calibrated
1	16,000	140	0.50	37,800	37,885–151,542
2	18,000	140	0.48	40,600	36,515–146,060
3	17,000	72	0.26	20,160	10,195–40,781
4	18,500	78	1.03	24,180	43,345–173,383
5	16,000	88	0.56	29,920	26,432–105,729
6	15,000	67	0.44	27,470	15,880–63,518
7	15,000	70	0.48	23,800	18,125–72,498
8	15,000	36	0.89	7200	17,261–69,045

2.2.3. Numerical Simulations

A three-dimensional finite difference-based groundwater flow model from MODFLOW-2005 [55] was used in the pre- and post-processing modelling environment of Processing MODFLOW [56] to simulate the saturated subsurface flow and surface water exchanges via the river–aquifer interface. In the present study, surface water leakage from the intermittent river was considered the main source of recharge for the aquifer [48]. The riverbed conductance (C, L^2T^{-1}) is a key parameter that controls the interaction between the surface water and groundwater and is represented in MODFLOW [55,57] by the following equation:

$$C = D_{riv}L_r k_0 / m_0 \tag{1}$$

where D_{riv} is the length of the river reach within the grid cell (L), L_r is the river width (L), m_0 is the thickness of the riverbed (L), and k_0 is the hydraulic conductivity of the riverbed (LT^{-1}).

The Streamflow Routing (STR) package [58] was selected to simulate the interactions between the river water and groundwater. This package was able to simulate the major features of the surface water in this study, i.e., changes in the flow along the river due to interactions with groundwater. In the STR package, the flow in a stream is instantaneously routed downstream. The streamflow routing is designed through a network of streams and always flows in the same direction along the streams.

The stream stages (H_r, L) of a rectangular stream channel are calculated using Manning's equation as follows [56]:

$$H_r = \left(\frac{Q \cdot n}{L_r \cdot S_{riv}^{\frac{1}{2}}}\right)^{\frac{3}{5}}$$

(2)

where Q is the calculated river discharge (L^3/T), S_{riv} is the slope of the river channel (L/L), and n is Manning's roughness coefficient (-).

The amount of recharge from precipitation was insignificant, and the total precipitation over the period of 2010–2012 was only 91 mm, based on observations at the local metrological station. For arid regions in north-western China, the direct recharging of rain-fed groundwater was estimated by the chloride mass balance method to be 1.5% of the mean annual precipitation [59,60]. In the study area, precipitation infiltration is most likely negligible, as indicated by the fact that single rainfall events of more than 5 mm were extremely rare between 2010 and 2012 and occurred predominantly during summer and autumn, when the potential evapotranspiration (PET) was extremely high (Figure 3). The PET was obtained from the water surface evaporation data observed by an E-601 evaporator at the local meteorological station (Figure 1) during the non-freezing period (April to October) and the calculated PET from Du, Yu, Wang, and Zhang [38] during the freezing period (November to the following March).

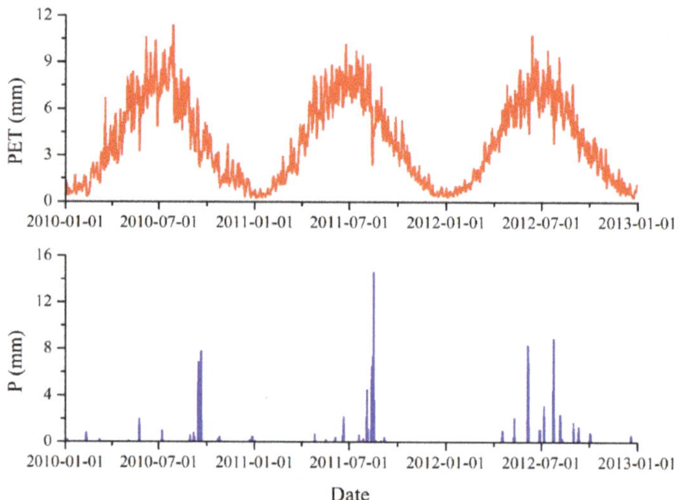

Figure 3. Daily potential evapotranspiration (PET, mm) and precipitation (P, mm) at the Ejina meteorological station.

Groundwater ET is a predominant discharge term in the water budget in the study area [39]. The depth of the water table in this area is mostly between 2 and 4 m [45]; therefore, the relationship between ET and water table depth can be simply assumed to be linear [61] with an extinction depth of 5 m [62]. For this reason, the Evapotranspiration (EVT) package [55], which assumes a linear relationship between ET and water table depth, was used to simulate the processes of ET. As shown in Figure 1, excluding the dominant landscape of the Gobi Desert, there is a large area of riparian oasis within the model domain. To simulate plant transpiration in the riparian oasis and soil evaporation in the Gobi Desert, we assigned different maximum ET rates (ET_{max}) for these two landscape types. The monthly ET_{max} values for the riparian oasis and Gobi Desert were independently calibrated to match the linear relationship between ET and water table depth within the interval of 2–4 m, while the monthly variations followed the monthly measured and calculated PET rates at the Ejina meteorological station (Figure 3).

Groundwater extraction by pumping wells for agricultural irrigation occurs primarily in the southern and north-eastern areas of the study region (Figure 1). Based on field investigations, the farmland area irrigated by a single irrigation well was approximately 1.08×10^5 m^2, and the daily water pumping rate of a farmland irrigation well was approximately 300 m^3/day during the irrigation period from May to August. The fraction of water returned to the aquifer as a result of agricultural irrigation was estimated to be approximately 0.1 [43]. Therefore, the water withdrawn from the aquifer for irrigation was 625 m^3/day within a grid cell of 2.50×10^5 m^2. The Well package implemented in MODFLOW-2005 [55] was used to simulate this groundwater withdrawal from the aquifer. The total number of cells containing irrigation wells within the simulation domain was 182.

Previous studies indicated that vertical unsaturated flow is insignificant in the Gobi Desert based on the water exchange between rivers and aquifers [63,64]. In addition, the dominant vegetation in the riparian area comprises groundwater-dependent species (e.g., *Populus* and *Tamarix*), which mostly use groundwater for transpiration [50]. Therefore, unsaturated flow in the study area can be neglected and was not addressed in the simulations.

The numerical model consisted of 225 rows and 41 columns, with a total simulation domain of 112.5×20.5 km. The simulation period was from 21 April 2010 to 7 September 2012, and the temporal resolution was 1 day. We used the measured daily river flow at the LHS from 1 January to 21 April 2010 and ran the model to obtain the initial groundwater levels. The simulated groundwater levels on 21 April 2010 were set as the initial groundwater levels (Figure 2).

2.2.4. Model Calibration

We used the combination of the measured river water (R1–R4) and groundwater (G1–G5) levels to calibrate the hydraulic conductivity, specific yield of the aquifer, riverbed hydraulic conductance, and GHB hydraulic conductance. The root mean squared error (*RMSE*, m) and correlation coefficient (*R*) were used to evaluate the goodness of fit of the observed and calculated levels [65]. The calibrated parameters of the aquifer and streambed are listed in Tables 1 and 2, respectively. Notably, the river in this study is intermittent with a wide riverbed, and the river width is highly dependent on the river stage. To account for the influence of the river width on the river–aquifer exchange, we used the temporal variations in the riverbed hydraulic conductance (see Table 2) associated with each river flow event in the simulations. In addition, the GHB hydraulic conductance was also calibrated, with values of 2000 m^2/day for the southern boundary and 2200 m^2/day for the northern boundary.

As shown in Figure 4, the simulated and observed water levels at the nine monitoring wells are generally consistent. The simulated water levels reflect the variations in the water levels at the observation points. The differences between the calculated and observed water levels are less than 0.5 m. The *RMSE* varies from 0.04 to 0.31, depending on the observation well. The correlation coefficients between the simulated and observed water levels at most monitoring wells are relatively high (0.77–0.95), with the exception of that at well G2, where no correlation is observed. This finding can be explained by the relatively large distance of well G2 from the Donghe River (Figure 2). Thus, the groundwater level at well G2 is less affected by the fluctuations in the river water level compared to the levels at the other monitoring wells, which are located closer to the river. The slight seasonal variations in the groundwater level at well G2 are more likely affected by the ET process. Subsequent research should aim to address simulating ET processes using a nonlinear relationship between ET_{max} and the groundwater level depth (e.g., [61,62]) to enhance the conformity of the observed and calculated levels in areas far from rivers.

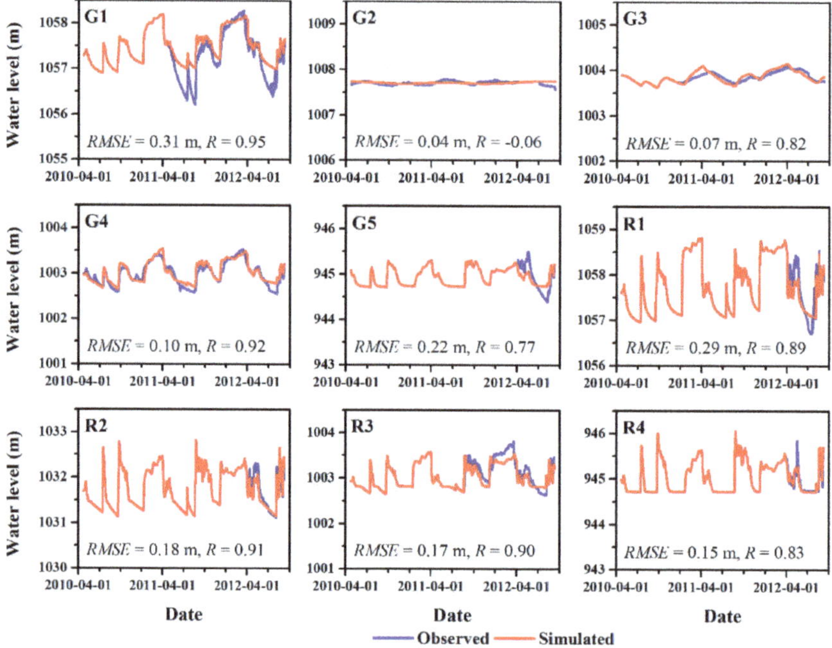

Figure 4. Simulated versus observed water levels from 21 April 2010 to 7 September 2012.

3. Results

3.1. Groundwater Budget

During the period from 21 April 2010 to 7 September 2012, the total groundwater recharge within the simulated region was 4.26×10^8 m^3. The river leakage through the streambed to the aquifer was 3.59×10^8 m^3, which accounted for approximately 84% of the total groundwater recharge. The other 16% of the groundwater recharge was due to lateral flow through the southern and northern model boundaries (6.77×10^7 m^3), which can be evidenced by hydro-geochemical analyses [48,54] and regional groundwater flow simulations [42].

The total groundwater discharge within the simulated region was 2.8×10^8 m^3, and the groundwater ET was 1.64×10^8 m^3, which accounted for 59% of the total discharge. Groundwater discharge to the river (5.53×10^7 m^3), groundwater exploitation (4.19×10^7 m^3), and water outflow through the southern and northern model boundaries (1.85×10^7 m^3) accounted for approximately 20%, 15%, and 6% of the total discharge, respectively.

The groundwater budget analysis indicated that river water leakage was the main recharge source, and that groundwater ET was the predominant discharge type (Table 3). As noted by Wang et al. [66] and Wang et al. [67], these two processes determine the changes in water storage and control the spatial and temporal dynamics of the studied groundwater system. The difference between the total recharge and discharge (1.46×10^8 m^3) can be explained by the increased groundwater storage in the model domain from 21 April 2010 to 7 September 2012. These results were consistent with previous studies (e.g., [39,45]), which demonstrated that, especially in the riparian zone, the groundwater level increased as a result of environmental flow controls aimed at delivering a set amount of surface water to the study area after 2000.

Table 3. Groundwater budget of the model domain from 21 April 2010 to 7 September 2012.

No.	Budget Component	Recharge		Discharge	
		Volume, m^3	%	Volume, m^3	%
1	Water exchange with river	3.59×10^8	84	5.53×10^7	20
2	Evapotranspiration	-	-	1.64×10^8	59
3	Groundwater exploitation	-		4.19×10^7	15
4	Lateral flow	6.77×10^7	16	1.85×10^7	6
5	Total	4.26×10^8	100	2.80×10^8	100

As shown in Figure 1, oasis area accounts for approximately 20% of the total simulation area, and the other 80% represents the Gobi Desert, where groundwater ET processes are negligible [63]. The growing season lasts from May to September and accounts for approximately 150 days per year. As a result, the average ET rate in the oasis area during the growing season is approximately 0.79 mm/day, which is close to *Tamarix*'s ET rate of 0.63–0.73 mm/day, as estimated using water table fluctuation methods during the growing seasons of 2010–2012 in this area [68].

3.2. Daily River Leakage

Daily river leakage through the streambed was highly dependent on river inflow, and river leakage generally followed river runoff processes (Figure 5). As shown in Figure 5, after high-flow events, e.g., from 18 August 2011 to 3 November 2011, the river leakage through the streambed was negative, which indicated that groundwater was discharged to the river. This finding can be explained by the fact that the riverbank stored surface water during the high stage, and when the river flow decreased rapidly, the groundwater stored in the riverbank was released into the river channel.

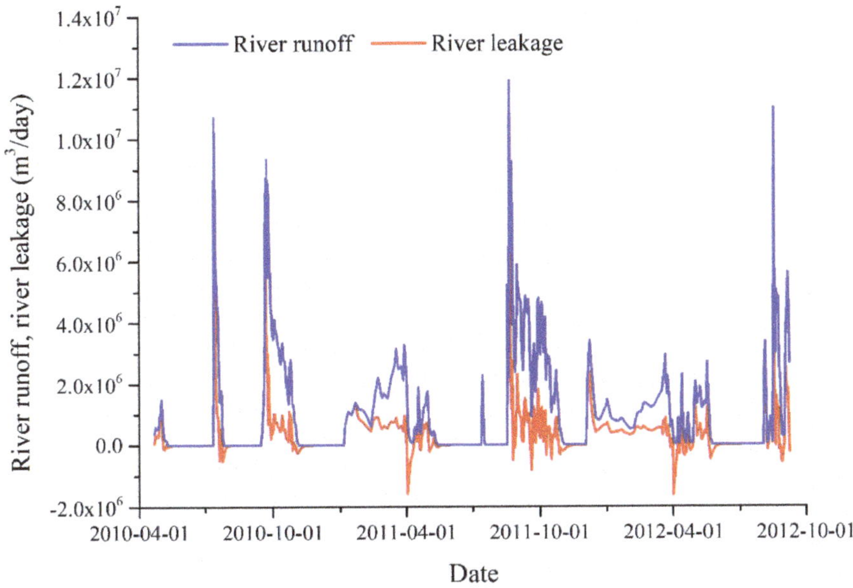

Figure 5. Observed daily runoff of the Donghe River at the LHS and simulated daily river leakage from 21 April 2010 to 7 September 2012.

As the total runoff of the Donghe River at the LHS was 9.19×10^8 m^3, the river leakage rate, i.e., the percentage of river leakage divided by the total inflow, in the river reaches from the LHS to the

Angcizha was approximately 33% from 21 April 2010 to 7 September 2012. However, the average leakage rates during the eight individual flow events varied significantly from 24% to 99.8% (Table 4).

Table 4. Leakage rates of the river flow periods during the simulation period; correlation coefficients indicate relationships between the stream runoff and river leakage.

River Flow Period	Total Runoff (10^6 m^3)	Average River Discharge (m^3/s)	River Leakage (10^6 m^3)	River Leakage Rate (%)	Correlation Coefficient (-)
21 April–9 May 2010	11.0	6.7	4.6	42	0.85
12–26 July 2010	50.7	39.1	20.3	40	0.79
15 September–4 November 2010	154.1	35.0	43.4	28	0.52
6 January–13 May 2011	166.8	15.1	65.5	39	0.25
13–17 July 2011	5.12	11.8	5.11	99.8	0.99
18 August–3 November 2011	248.7	36.9	59.1	24	0.60
4 December 2011–24 May 2012	192.0	12.8	73	38	0.61
3 August–7 September 2012	91.5	29.4	36.2	40	0.57

For events with short-term flow durations (e.g., from 21 April to 9 May 2010 and 13 to 17 July 2011), the river leakage rates were relatively high, with values of 42–99.8%. In contrast, for events with long-term flow durations (e.g., from 18 August to 3 November 2011 and 4 December 2011 to 24 May 2012), the river leakage rates were relatively low, with values of 24–38%. This difference is mainly associated with the decreasing hydraulic gradient between the surface water and groundwater during events with long-term flow durations.

Table 4 shows the correlation coefficients between the stream leakage and river runoff for eight individual flow events. These correlation coefficients are relatively high (from 0.52 to nearly 1, except for the flow event from 6 January to 13 May 2011 with a value of 0.25). In addition, there is a common tendency that shorter flow events during summer periods exhibit higher correlations. For example, short flow events, e.g., 13–17 July 2011, 21 April–9 May 2010, and 12–26 July 2010, have correlation coefficients of 0.79–0.99, while relatively long flow events, e.g., 6 January–13 May 2011, 3 August–7 September 2012, and 4 December 2011–24 May 2012, have correlation coefficients of 0.25–0.61. Therefore, it can be stated that if water needs to be transported from Langxinshan to terminal lakes with minimum leakage, the water should be transported during winter periods. However, to increase the groundwater level in the riparian zone and support riparian ecosystems, it is better to transport water predominantly during spring and summer periods.

From the perspective of ground and surface water interactions, the interaction regime is constantly connected, which means that even after continuous periods without runoff (for example, from April to June), the groundwater level remains above the bottom of the riverbed sediments. Thus, water from the stream percolates via saturated sediments, which contributes to relatively high leakage rates.

3.3. Parameter Sensitivity Analysis

To investigate the influence of the model parameters on the simulation results, an uncertainty analysis was conducted. The present study revealed the model sensitivity by analyzing the effects of the ET rate, aquifer parameters, and streambed parameters on the cumulative river leakage.

Sensitivity represents the effect of variations in one parameter on the calculation results and is generally represented by the following equation [69]:

$$\beta_k = \frac{\Delta Q}{\Delta \alpha_k} \approx \frac{Q(\alpha_k + \Delta \alpha_k) - Q(\alpha_k)}{\Delta \alpha_k} \tag{3}$$

where β_k represents the sensitivity of the model variable (leakage (Q) in the present study) to the parameter, α_k represents the actual parameter value, $\Delta \alpha_k$ represents the change in the parameter value,

$Q(\alpha_k + \Delta\alpha_k)$ represents the leakage simulated by the model after the parameter variation and $Q(\alpha_k)$ represents the leakage simulated by the model before the parameter variation.

A local sensitivity analysis method was used to analyze the sensitivity of the model parameters to river leakage. Specifically, we changed the hydraulic conductivity (K) and specific yield (S_y) of the aquifer, the maximum ET rate (ET_{max}), and the riverbed conductance (C) by −20% to 20%. The parameter range selected for the sensitivity analysis was based on in situ experiments of streambed hydraulic conductivity [13,16] and the observed potential evaporation in the study area (Figure 3). Only one parameter was changed each time to determine the leakage variation (Figure 6). The results of the sensitivity analysis showed that the hydraulic conductivity of the aquifer had the greatest effect on leakage. When the hydraulic conductivity changed by 20%, the leakage changed by approximately 11%. Thus, the sensitivity of leakage to the hydraulic conductivity of the aquifer was 0.57. The calculations indicated that the sensitivities of leakage to the maximum ET rate, specific yield, and riverbed conductance were 0.28, 0.08, and 0.02, respectively. Therefore, leakage was most sensitive to the hydraulic conductivity of the aquifer, followed by the maximum ET rate, specific yield of the aquifer, and riverbed conductance.

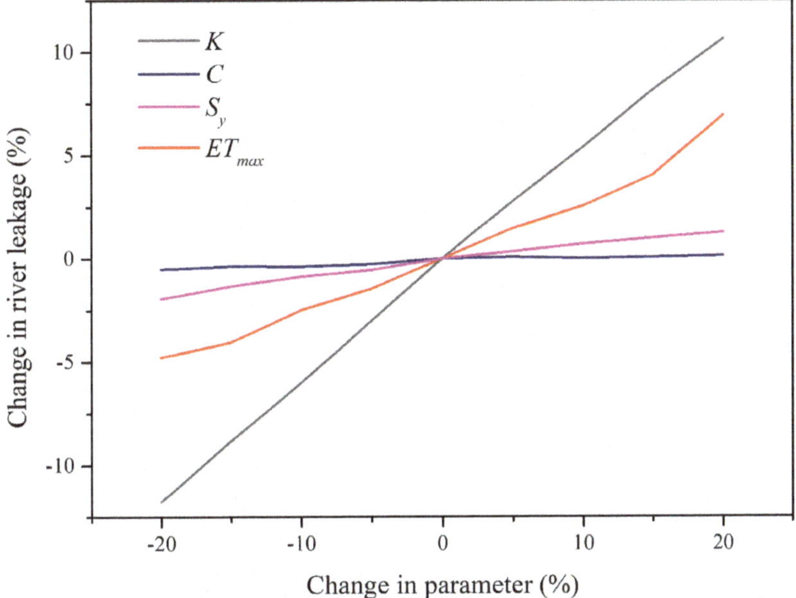

Figure 6. Sensitivity of river leakage to the model parameters.

The significant sensitivity of the river leakage to the maximum ET rate can be explained by the increase in the hydraulic gradient between the surface and groundwater when ET_{max} rises, which causes a decline in the groundwater level. Figure 7 demonstrates that if we enhance ET_{max}, the hydraulic gradient rises, causing the leakage rate from the river to increase. Thus, according to the simulation results, riverbank recharge processes are highly dependent on riparian ET processes.

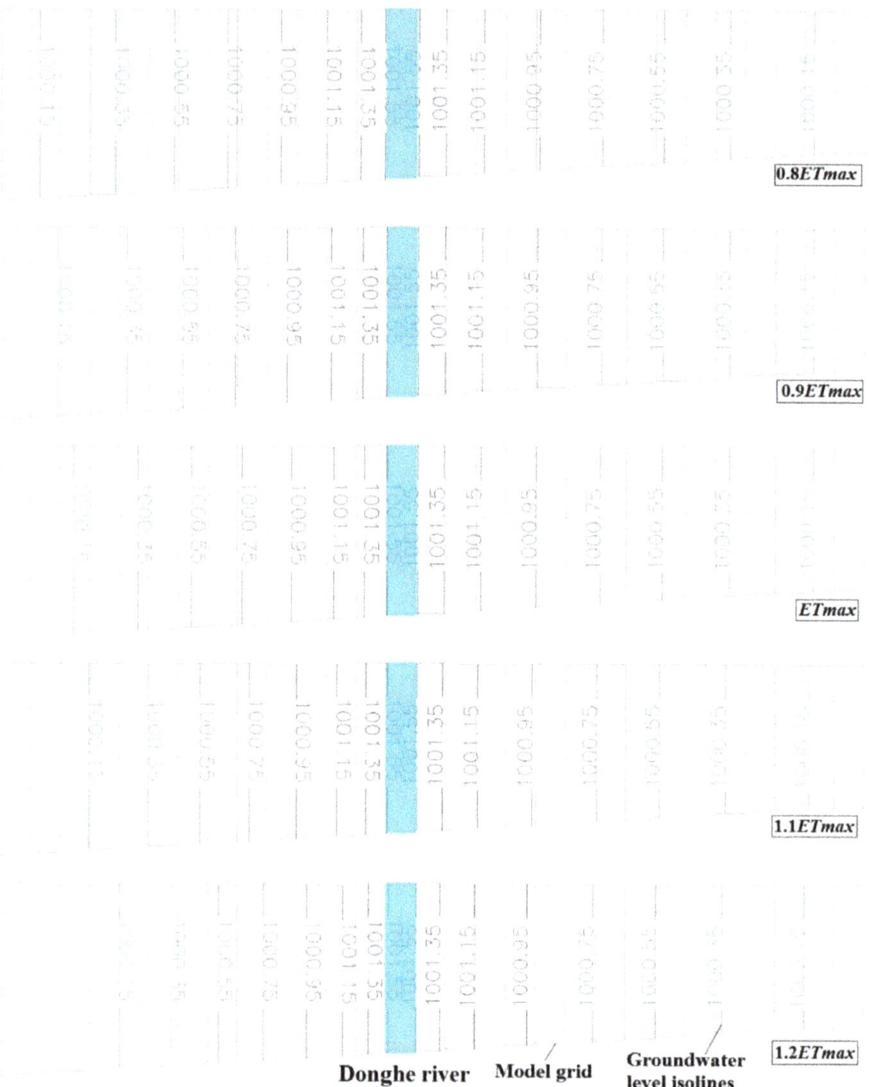

Figure 7. Groundwater levels under different values of the maximum evapotranspiration rate (cross-section from west to east in the middle of the simulated area).

4. Discussion

To explain the high sensitivity of river losses to groundwater ET in the riparian area, we consider a supporting analytical model for the formation of groundwater flow from the river towards the ET zone in the riparian area adjacent to the river channel. For this purpose, let us consider the one-dimensional groundwater flow formed by losses from the river and discharged by the ET of groundwater in the adjacent riparian zone. The details of the formulation of the corresponding groundwater flow boundary problem are given in Appendix A. As shown in Appendix A, water losses from the river during stationary groundwater flow periods are controlled by two hydraulic parameters expressed in length units. The first parameter, which is the streambed hydraulic resistance length, ΔL, characterizes the additional hydraulic resistance due to the bottom sediments of the river channel [36]. The length ΔL is expressed as follows:

$$\Delta L = b_r^{-1} \coth(0.5 L_r b_r); \; b_r = \sqrt{\frac{k_0}{m_0 T}} \tag{4}$$

The second parameter is the characteristic width, L_{ET}, of the groundwater ET zone adjacent to the river channel. This length depends on the parameters characterizing the decrease in ET with the depth of the groundwater table and is expressed as follows:

$$L_{ET} = \sqrt{\frac{d_{max} T}{ET_{max}}}; \; d_{max} = Z_{surf} - Z_{crit} \tag{5}$$

The expression for the specific flow rate of losses from the river, q, to the riparian area is as follows:

$$q = 2T\frac{H_r - Z_{crit}}{L_{ET} + \Delta L} = 2T\frac{H_r - Z_{crit}}{L_{ET}(1 + \alpha_r)}; \; \alpha_r = \sqrt{\frac{ET_{max} m_0}{d_{max} k_0}} \coth(0.5 L_r b_r) \tag{6}$$

The factor of 2 in Equation (6) assumes that there will be symmetrical losses to both riverbanks with riparian zones. The notations for Equations (4)–(6) are listed in the Appendix A.

The equations shown above help to clarify the fact that the sensitivity of leakage to riverbed conductance is only 0.02, while that to the hydraulic conductivity of the aquifer is 0.57, which can be explained by the value of the streambed hydraulic resistance length (ΔL), which is between 60 and 80 m, depending on the number of river segments (Table 2). The parameter ΔL characterizes the hydraulic resistance of the bottom sediments (Appendix A). In the study area, the hydraulic conductivity of the riverbed sediments is relatively high, i.e., 1–40 m/day [41], which leads to relatively small bottom sediment hydraulic resistance values.

In addition, an analytical criterion, α_r (Appendix A), can be calculated using typical values of ET_{max} during the vegetation period as a function of k_0, m_0, and d_{max} (extinction depth). The calculated value of α_r is 0.01, which indicates almost no sensitivity of leakage to the streambed hydraulic conductance. However, the sensitivities of leakage to the hydraulic conductivity of the aquifer and ET_{max} are comparable. Thus, under the present conditions, river leakage is more sensitive to the hydraulic parameters related to the aquifer and ET rather than to the riverbed, and thus, it is more important to study aquifer parameters and ET parameters to accurately estimate river leakage from such intermittent streams under arid conditions. The present conclusion is valid for cases with high maximum ET rate values and relatively low ΔL values.

Equally important is that the EVT package [55] selected in this study to simulate ET processes is oversimplified and is highly based on previous empirical analysis of the dependence of ET on the water table depth between 2 and 4 m [45,68]. This approach is probably acceptable for analyses of surface–groundwater exchanges at regional scales. However, to address riparian ET processes, nonlinear ET models that account for plant types with different rooting depths (e.g., [70]), or even physical-based models with dynamic root optimization schemes (e.g., [71]), are required.

5. Conclusions

The present study conducted coupled simulations of surface water and groundwater exchanges in an arid ephemeral stream–aquifer system using the MODFLOW-2005 code with the STR package. The results showed that from 21 April 2010 to 7 September 2012, river water leakage accounted for 33% of the total river runoff for the simulated segments.

A sensitivity analysis demonstrated that the most important parameters of the studied system that influence river leakage are the hydraulic conductance of the aquifer and the maximum ET rate. Almost no sensitivity was obtained for the riverbed hydraulic conductance, which was explained by the relatively high hydraulic conductivity of the riverbed sediments. Thus, instead of studying the hydraulic parameters of riverbeds, further research should focus on studying the ET parameters and selecting an appropriate ET model that reflects the eco-physiology of riparian ecosystems [70]. The present conclusion is valid only for cases with relatively high streambed hydraulic conductivities (compared to the hydraulic conductivity of the aquifer) in arid regions, as demonstrated in the studied case. Coupled surface/groundwater models, which are used to estimate river leakage, should consider riparian zones because they play a dominant role in the formation of leakage from rivers for recharging via ET.

As the model synchronously simulated the daily variation in the river water and groundwater levels that affected leakage, the simulation results are more reliable than those of previous models that used only groundwater level data collected over long periods for verification when simulating leakage (temporal resolution greater than 10 days). To the best of our knowledge, the present study is the first to simulate and analyze the daily river leakage process under the conditions of ecological water transport in the downstream region of the Heihe River. The study results can provide scientific evidence for further ecological water transport research.

Author Contributions: Conceptualization, J.Y., P.W., and S.P.P.; methodology, L.M., P.Y.V., and S.P.P.; validation, P.W.; data curation, L.M. and P.W.; writing—original draft preparation, L.M., and P.W.; writing—review and editing, P.Y.V., P.W., and S.P.P.; supervision, J.Y.; and funding acquisition, P.W., J.Y., P.Y.V., and S.P.P. All authors have read and agreed to the published version of the manuscript.

Funding: This research was supported by grants from the National Natural Science Foundation of China (Nos. 41671023 and 41877165) and the NSFC-RFBR Programme 2018–2019 (Nos. 41811530084 and 18-55-53025 ГФЕН_a), and the second author was supported by the Russian Foundation for Basic Research (No. 19-35-90014 Аспиранты).

Acknowledgments: We would like to thank Grinevsky Sergey of the Department of Hydrogeology, Lomonosov Moscow State University, for his helpful advice and Juntao Zhu, Yongliang Xu, Zhiyong Wang, and Dandan Wang for their participation in the fieldwork. Ping Wang and Sergey P. Pozdniakov are grateful for support from the Special Exchange Program of the Chinese Academy of Sciences 2019–2020. The authors also gratefully acknowledge the anonymous reviewers for their valuable comments and suggestions that have led to substantial improvements over an earlier version of the manuscript.

Appendix A

Groundwater flow from a river to a riparian evapotranspiration (ET) zone.

In an arid region with a groundwater ET zone located along a stream in summer (Figure A1), river water infiltrates the aquifer. This water is discharged within the riparian zone with shallow groundwater levels. Consider the situation shown in Figure A1, and suppose that the groundwater flow consists of two zones.

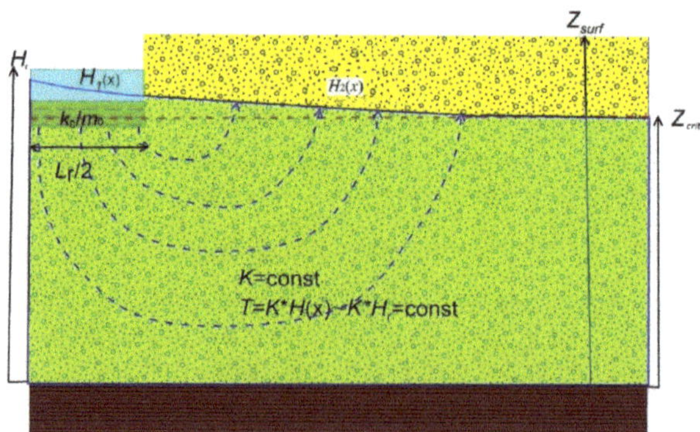

Figure A1. Schematic flow of groundwater discharge in an evapotranspiration (ET) zone located along a stream.

Zone 1 is the zone below the stream. In this zone, the surface–groundwater exchange rate, $v_z(x)$, is proportional to the head difference between the river, H_r, and groundwater, $H_1(x)$, as well as the hydraulic resistance of bottom sediments as follows:

$$v_z(x) = \frac{k_0}{m_0}(H_r - H_1(x)) \tag{A1}$$

where k_0 and m_0 are the hydraulic conductivity and the thickness of the bottom sediments, respectively, and x is the distance perpendicular to the river direction.

In the next zone, zone 2, the rate of groundwater discharge $ET_{gw}(x)$ can be described as a linear function of the groundwater depth as follows:

$$ET_{gw}(x) = ET_{max}\left(\frac{Z_{surf} - H_2(x)}{d_{max}}\right), \ 0 \leq Z_{surf} - H_2(x) \leq d_{max} \tag{A2}$$

where ET_{max} is the PET and d_{max} is the groundwater ET extinction depth.

Let us consider a steady-state groundwater flow such that the groundwater recharge from the river is balanced by the groundwater ET. Using the Dupuit precondition regarding the head hydrostatic distribution of the aquifer saturated thickness and constant transmissivity, we can express the following system of 1D steady-state equations for the groundwater flow:

$$\frac{d^2H_1}{dx^2} - b_r^2(H_1 - H_r) = 0, \ \frac{d^2H_2}{dx^2} - b_{ET}^2(H_2 - H_{crit}) = 0, b_r = \sqrt{\frac{k_0}{m_0 T}}; \ b_{ET} = \sqrt{\frac{ET_{max}}{d_{max}T}};$$
$$d_{max} = Z_{surf} - Z_{crit}, \tag{A3}$$

where T is the transmissivity of the aquifer, and Z_{crit} is the absolute position of the groundwater ET extinction depth.

The boundary conditions for this system assume symmetry of flow in the middle of the river and equality of heads and gradients at the boundaries of the zones as follows:

$$\begin{aligned} x = 0 &: \frac{dH_1}{dx} = 0 \\ x = \tfrac{1}{2}L_r &: H_1 = H_2; \frac{dH_1}{dx} = \frac{dH_2}{dx} \\ X \to \infty &: H_2 = Z_{crit} \end{aligned} \tag{A4}$$

The general solution of system (Equation (A3)) is:

$$\begin{aligned} H_r - H_1(x) &= C_1 \exp(-b_r x) + C_2 \exp(b_r x) \\ H_2(x) - Z_{crit} &= C_3 \exp(-b_e x) + C_4 \exp(b_e x) \end{aligned} \tag{A5}$$

The unknown coefficients can be determined using (Equation (A4)) as follows:

$$\begin{aligned} H_1(x) &= H_r - (H_r - Z_{crit})b_{ET}\frac{\cosh(b_r x)}{b_{ET}\cosh(0.5L_r b_r) + b_r \sinh(0.5L_r b_r)} \\ H_2(x) &= Z_{crit} + (H_r - Z_{crit})b_r\frac{\exp(-b_{ET}x)}{\exp(-0.5L_r b_{ET})}\frac{\sinh(0.5L_r b_r)}{b_{ET}\cosh(0.5L_r b_r) + b_r \sinh(0.5L_r b_r)} \end{aligned} \tag{A6}$$

Consider the specific flow q from the river to the riparian zone as follows:

$$q = -T\frac{dH_1}{dx}\Big|_{x=0.5L_r} = T\frac{H_r - Z_{crit}}{\Delta L + L_{ET}} \tag{A7}$$
$$\Delta L = b_r^{-1}\coth(0.5L_r b_r); \ L_{ET} = b_{ET}^{-1}$$

For the more general case of a limited riparian zone with a width of $0.5L_0$, note that (Equation (A7)) is also valid, although L_{ET} should be calculated as follows:

$$L_{ET} = b_{ET}^{-1}\coth(0.5L_0 b_{ET}) \tag{A8}$$

Thus, the groundwater flow from the river used for ET depends on two characteristic lengths. The first length, ΔL, characterizes the hydraulic resistance of the bottom sediments. The second length, L_{ET}, characterizes the hydraulic resistance of the groundwater discharge due to evaporation, which changes linearly with the depth of the groundwater table. Thus, (Equation (A7)) can be rewritten as follows:

$$q = T\frac{H_r - Z_{crit}}{L_{ET}(1 + \alpha_r)}; \alpha_r = \sqrt{\frac{ET_{max}m_0}{d_{max}k_0}}\coth(0.5L_rb_r) \tag{A9}$$

For a small α_r ($\alpha_r < 0.1$), the hydraulic resistance of the bottom sediments does not play an important role in the river water loss due to ET, while for a large α_r ($\alpha_r > 10$), the losses of river water due to ET are controlled by the hydraulic resistance of the bottom sediments.

References

1. Wang, P. Progress and prospect of research on water exchange between intermittent rivers and aquifers in arid regions of northwestern China. *Prog. Geogr.* **2018**, *37*, 183–197, (In Chinese with English abstract).
2. Datry, T.; Larned, S.T.; Tockner, K. Intermittent Rivers: A Challenge for Freshwater Ecology. *BioScience* **2014**, *64*, 229–235. [CrossRef]
3. Sophocleous, M. Interactions between groundwater and surface water: The state of the science. *Hydrogeol. J.* **2002**, *10*, 52–67. [CrossRef]
4. Brunner, P.; Therrien, R.; Renard, P.; Simmons, C.T.; Franssen, H.-J.H. Advances in understanding river-groundwater interactions. *Rev. Geophys.* **2017**, *55*, 818–854. [CrossRef]
5. Filimonova, E.A.; Baldenkov, M.G. A combined-water-system approach for tackling water scarcity: Application to the Permilovo groundwater basin, Russia. *Hydrogeol. J.* **2016**, *24*, 489–502. [CrossRef]
6. Pekel, J.-F.; Cottam, A.; Gorelick, N.; Belward, A.S. High-resolution mapping of global surface water and its long-term changes. *Nature* **2016**, *540*, 418. [CrossRef]
7. Partington, D.; Therrien, R.; Simmons, C.T.; Brunner, P. Blueprint for a coupled model of sedimentology, hydrology, and hydrogeology in streambeds. *Rev. Geophys.* **2017**, *55*, 287–309. [CrossRef]
8. Constantz, J. Streambeds merit recognition as a scientific discipline. *Wiley Interdiscip. Rev. Water* **2016**, *3*, 13–18. [CrossRef]
9. Filimonova, E.; Shtengelov, R. The dependence of stream depletion by seasonal pumping on various hydraulic characteristics and engineering factors. *Hydrogeol. J.* **2013**, *21*, 1821–1832. [CrossRef]
10. Song, J.; Chen, X.; Cheng, C.; Summerside, S.; Wen, F. Effects of hyporheic processes on streambed vertical hydraulic conductivity in three rivers of Nebraska. *Geophys. Res. Lett.* **2007**, *34*, L07409. [CrossRef]
11. Xian, Y.; Jin, M.; Zhan, H. Buffer effect on identifying transient streambed hydraulic conductivity with inversion of flood wave responses. *J. Hydrol.* **2020**, *580*, 124261. [CrossRef]
12. Tang, Q.; Kurtz, W.; Schilling, O.S.; Brunner, P.; Vereecken, H.; Hendricks Franssen, H.J. The influence of riverbed heterogeneity patterns on river-aquifer exchange fluxes under different connection regimes. *J. Hydrol.* **2017**, *554*, 383–396. [CrossRef]
13. Min, L.; Yu, J.; Liu, C.; Zhu, J.; Wang, P. The spatial variability of streambed vertical hydraulic conductivity in an intermittent river, northwestern China. *Environ. Earth Sci.* **2013**, *69*, 873–883. [CrossRef]
14. Chen, X.; Song, J.; Wang, W. Spatial variability of specific yield and vertical hydraulic conductivity in a highly permeable alluvial aquifer. *J. Hydrol.* **2010**, *388*, 379–388. [CrossRef]
15. Wu, G.; Shu, L.; Lu, C.; Chen, X.; Zhang, X.; Appiah-Adjei, E.; Zhu, J. Variations of streambed vertical hydraulic conductivity before and after a flood season. *Hydrogeol. J.* **2015**, *23*, 1603–1615. [CrossRef]
16. Wang, P.; Pozdniakov, S.P.; Vasilevskiy, P.Y. Estimating groundwater-ephemeral stream exchange in hyper-arid environments: Field experiments and numerical simulations. *J. Hydrol.* **2017**, *555*, 68–79. [CrossRef]
17. Hatch, C.E.; Fisher, A.T.; Revenaugh, J.S.; Constantz, J.; Ruehl, C. Quantifying surface water-groundwater interactions using time series analysis of streambed thermal records: Method development. *Water Resour. Res.* **2006**, *42*, W10410. [CrossRef]
18. Weber, M.D.; Booth, E.G.; Loheide, S.P. Dynamic ice formation in channels as a driver for stream-aquifer interactions. *Geophys. Res. Lett.* **2013**, *40*, 3408–3412. [CrossRef]

19. Tooth, S. Process, form and change in dryland rivers: A review of recent research. *Earth Sci. Rev.* **2000**, *51*, 67–107. [CrossRef]

20. Xian, Y.; Jin, M.; Liu, Y.; Si, A. Impact of lateral flow on the transition from connected to disconnected stream–aquifer systems. *J. Hydrol.* **2017**, *548*, 353–367. [CrossRef]

21. Phogat, V.; Potter, N.J.; Cox, J.W.; Šimůnek, J. Long-Term Quantification of Stream-Aquifer Exchange in a Variably-Saturated Heterogeneous Environment. *Water Resour. Manag.* **2017**, *31*, 4353–4366. [CrossRef]

22. Brunner, P.; Cook, P.G.; Simmons, C.T. Disconnected Surface Water and Groundwater: From Theory to Practice. *Ground Water* **2011**, *49*, 460–467. [CrossRef]

23. Brunner, P.; Cook, P.G.; Simmons, C.T. Hydrogeologic controls on disconnection between surface water and groundwater. *Water Resour. Res.* **2009**, *45*, W01422. [CrossRef]

24. Cook, P.G. Quantifying river gain and loss at regional scales. *J. Hydrol.* **2015**, *531 Pt 3*, 749–758. [CrossRef]

25. Kalbus, E.; Reinstorf, F.; Schirmer, M. Measuring methods for groundwater-surface water interactions: A review. *Hydrol. Earth Syst. Sci.* **2006**, *10*, 873–887. [CrossRef]

26. Landon, M.K.; Rus, D.L.; Harvey, F.E. Comparison of instream methods for measuring hydraulic conductivity in sandy streambeds. *Ground Water* **2001**, *39*, 870–885. [CrossRef]

27. Shanafield, M.; Cook, P.G. Transmission losses, infiltration and groundwater recharge through ephemeral and intermittent streambeds: A review of applied methods. *J. Hydrol.* **2014**, *511*, 518–529. [CrossRef]

28. Pozdniakov, S.P.; Wang, P.; Lekhov, M.V. A semi-analytical generalized Hvorslev formula for estimating riverbed hydraulic conductivity with an open-ended standpipe permeameter. *J. Hydrol.* **2016**, *540*, 736–743. [CrossRef]

29. Caissie, D.; Luce, C.H. Quantifying streambed advection and conduction heat fluxes. *Water Resour. Res.* **2017**, *53*, 1595–1624. [CrossRef]

30. Zhou, D.; Zhang, Y.; Gianni, G.; Lichtner, P.; Engelhardt, I. Numerical modelling of stream–aquifer interaction: Quantifying the impact of transient streambed permeability and aquifer heterogeneity. *Hydrol. Process.* **2018**, *32*, 2279–2292. [CrossRef]

31. Rau, G.C.; Halloran, L.J.S.; Cuthbert, M.O.; Andersen, M.S.; Acworth, R.I.; Tellam, J.H. Characterising the dynamics of surface water-groundwater interactions in intermittent and ephemeral streams using streambed thermal signatures. *Adv. Water Resour.* **2017**, *107*, 354–369. [CrossRef]

32. Batlle-Aguilar, J.; Cook, P.G. Transient infiltration from ephemeral streams: A field experiment at the reach scale. *Water Res. Res.* **2012**, *48*, W11518. [CrossRef]

33. Niswonger, R.G.; Prudic, D.E.; Fogg, G.E.; Stonestrom, D.A.; Buckland, E.M. Method for estimating spatially variable seepage loss and hydraulic conductivity in intermittent and ephemeral streams. *Water Resour. Res.* **2008**, *44*, 1–14. [CrossRef]

34. Reid, M.E.; Dreiss, S.J. Modeling the effects of unsaturated, stratified sediments on groundwater recharge from intermittent streams. *J. Hydrol.* **1990**, *114*, 149–174. [CrossRef]

35. Leake, S.A.; Gungle, B. *Evaluation of Simulations to Understand Effects of Groundwater Development and Artificial Recharge on the Surface Water and Riparian Vegetation Sierra Vista Subwatershed, Upper San Pedro Basin, Arizona*; 2012-1206; U.S. Geological Survey: Reston, VA, USA, 2012.

36. Wang, P.; Pozdniakov, S.P.; Shestakov, V.M. Optimum experimental design of a monitoring network for parameter identification at riverbank well fields. *J. Hydrol.* **2015**, *523*, 531–541. [CrossRef]

37. Doppler, T.; Franssen, H.-J.H.; Kaiser, H.-P.; Kuhlman, U.; Stauffer, F. Field evidence of a dynamic leakage coefficient for modelling river–aquifer interactions. *J. Hydrol.* **2007**, *347*, 177–187. [CrossRef]

38. Du, C.; Yu, J.; Wang, P.; Zhang, Y. Reference Evapotranspiration Changes: Sensitivities to and Contributions of Meteorological Factors in the Heihe River Basin of Northwestern China (1961–2013; 2014). *Adv. Meteorol.* **2016**, *2016*, 4143580. [CrossRef]

39. Wang, P.; Yu, J.; Pozdniakov, S.P.; Grinevsky, S.O.; Liu, C. Shallow groundwater dynamics and its driving forces in extremely arid areas: A case study of the lower Heihe River in northwestern China. *Hydrol. Process.* **2014**, *28*, 1539–1553. [CrossRef]

40. Liu, X.; Yu, J.; Wang, P.; Zhang, Y.; Du, C. Lake Evaporation in a Hyper-Arid Environment, Northwest of China—Measurement and Estimation. *Water* **2016**, *8*, 527. [CrossRef]

41. Vasilevskiy, P.Y.; Wang, P.; Pozdniakov, S.P.; Davis, P. Revisiting the modified Hvorslev formula to account for the dynamic process of streambed clogging: Field validation. *J. Hydrol.* **2019**, *568*, 862–866. [CrossRef]

42.   Yao, Y.; Zheng, C.; Liu, J.; Cao, G.; Xiao, H.; Li, H.; Li, W. Conceptual and numerical models for groundwater flow in an arid inland river basin. *Hydrol. Process.* **2015**, *29*, 1480–1492. [CrossRef]

43.   Xu, Y.; Yu, J.; Zhang, Y.; Wang, P.; Wang, D. Groundwater dynamic numerical simulation in the Ejina Oasis in an ecological water conveyance period. *Hydrogeol. Eng. Geol.* **2014**, *41*, 11–18, (In Chinese with English abstract).

44.   Akiyama, T.; Sakai, A.; Yamazaki, Y.; Wang, G.; Fujita, K.; Nakawo, M.; Masayoshi, A.; Jurhpei, A.; Yuki, G. Surface water-groundwater interaction in the Heihe River basin, Northwestern China. *Bull. Glaciol. Res.* **2007**, *24*, 87–94.

45.   Wang, P.; Yu, J.; Zhang, Y.; Fu, G.; Min, L.; Ao, F. Impacts of environmental flow controls on the water table and groundwater chemistry in the Ejina Delta, northwestern China. *Environ. Earth Sci.* **2011**, *64*, 15–24. [CrossRef]

46.   Wu, X.; Shi, S.; Li, Z.; Hao, A.; Qiao, W.; Yu, Z.; Zhang, S. The study on the groundwater flow system of Ejina basin in lower reaches of the Heihe River in Northwest China (Part 1). *Hydrogeol. Eng. Geol.* **2002**, *1*, 16–20, (In Chinese with English abstract).

47.   Xie, Q. *Regional Hydrogeological Survey Report of the People's Republic of China (1:200,000): Ejina K-47-[24] [R]*; Chinese People's Liberation Army: Jiuquan, China, 1980. (In Chinese)

48.   Wang, P.; Yu, J.; Zhang, Y.; Liu, C. Groundwater recharge and hydrogeochemical evolution in the Ejina Basin, northwest China. *J. Hydrol.* **2013**, *476*, 72–86. [CrossRef]

49.   Qin, D.; Zhao, Z.; Han, L.; Qian, Y.; Ou, L.; Wu, Z.; Wang, M. Determination of groundwater recharge regime and flowpath in the Lower Heihe River basin in an arid area of Northwest China by using environmental tracers: Implications for vegetation degradation in the Ejina Oasis. *Appl. Geochem.* **2012**, *27*, 1133–1145. [CrossRef]

50.   Wang, P.; Zhang, Y.; Yu, J.; Fu, G.; Ao, F. Vegetation dynamics induced by groundwater fluctuations in the lower Heihe River Basin, northwestern China. *J. Plant Ecol.* **2011**, *4*, 77–90. [CrossRef]

51.   Zhang, Y.; Yu, J.; Wang, P.; Fu, G. Vegetation responses to integrated water management in the Ejina basin, northwest China. *Hydrol. Process.* **2011**, *25*, 3448–3461. [CrossRef]

52.   Yao, Y.; Tian, Y.; Andrews, C.; Li, X.; Zheng, Y.; Zheng, C. Role of Groundwater in the Dryland Ecohydrological System: A Case Study of the Heihe River Basin. *J. Geophys. Res. Atmos.* **2018**, *123*, 6760–6776. [CrossRef]

53.   Wen, X.; Wu, Y.; Su, J.; Zhang, Y.; Liu, F. Hydrochemical characteristics and salinity of groundwater in the Ejina Basin, Northwestern China. *Environ. Geol.* **2005**, *48*, 665–675. [CrossRef]

54.   Yang, Q.; Xiao, H.; Zhao, L.; Yang, Y.; Li, C.; Zhao, L.; Yin, L. Hydrological and isotopic characterization of river water, groundwater, and groundwater recharge in the Heihe River basin, northwestern China. *Hydrol. Process.* **2011**, *25*, 1271–1283. [CrossRef]

55.   Harbaugh, A.W. *MODFLOW-2005, the U.S. Geological Survey Modular Ground-Water Model—The Ground-Water Flow Process*; U.S. Geological Survey Techniques and Methods 6-A16; U.S. Geological Survey: Reston, VA, USA, 2005.

56.   Chiang, W.-H. *3D-Groundwater Modeling with PMWIN: A Simulation System for Modeling Groundwater Flow and Transport Processes*, 2nd ed.; Springer: New York, NY, USA, 2005; p. 397.

57.   McDonald, M.G.; Harbaugh, A.W. *A Modular Three-Dimensional Finite-Difference Ground-Water Flow Model*; U.S. Geological Survey: Reston, VA, USA, 1988.

58.   Prudic, D.E. *Documentation of a Computer Program to Simulate Stream-Aquifer Relations Using a Modular, Finite-Difference, Ground-Water Flow Model*; U.S. Geological Survey: Reston, VA, USA, 1989; pp. 88–729.

59.   Gates, J.; Edmunds, W.; Ma, J.; Scanlon, B. Estimating groundwater recharge in a cold desert environment in northern China using chloride. *Hydrogeol. J.* **2008**, *16*, 893–910. [CrossRef]

60.   Zhu, G.F.; Li, Z.Z.; Su, Y.H.; Ma, J.Z.; Zhang, Y.Y. Hydrogeochemical and isotope evidence of groundwater evolution and recharge in Minqin Basin, Northwest China. *J. Hydrol.* **2007**, *333*, 239–251. [CrossRef]

61.   Hu, S.; Tian, C.; Song, Y.; Chen, X.; Li, Y. Models for calculating phreatic water evaporation on bare and Tamarix-vegetated lands. *Chin. Sci. Bull.* **2006**, *51*, 43–50. [CrossRef]

62.   Shah, N.; Nachabe, M.; Ross, M. Extinction depth and evapotranspiration from ground water under selected land covers. *Ground Water* **2007**, *45*, 329–338. [CrossRef]

63.   Du, C.; Yu, J.; Wang, P.; Zhang, Y. Analysing the mechanisms of soil water and vapour transport in the desert vadose zone of the extremely arid region of northern China. *J. Hydrol.* **2018**, *558*, 592–606. [CrossRef]

64. Guo, X.; Feng, Q.; Si, J.; Xi, H.; Zhao, Y.; Deo, R.C. Partitioning groundwater recharge sources in multiple aquifers system within a desert oasis environment: Implications for water resources management in endorheic basins. *J. Hydrol.* **2019**, *579*, 124212. [CrossRef]

65. Legates, D.R.; McCabe, G.J., Jr. Evaluating the use of "goodness-of-fit" measures in hydrologic and hydroclimatic model validation. *Water Resour. Res.* **1999**, *35*, 233–241. [CrossRef]

66. Wang, P.; Yu, J.; Min, L.; Xu, Y.; Zhu, B.; Zhang, Y.; Du, C. Shallow groundwater regime and its driving forces in the Elina oasis. *Quat. Sci.* **2014**, *34*, 982–993, (In Chinese with English abstract).

67. Wang, T.-Y.; Wang, P.; Zhang, Y.-C.; Yu, J.-J.; Du, C.-Y.; Fang, Y.-H. Contrasting groundwater depletion patterns induced by anthropogenic and climate-driven factors on Alxa Plateau, northwestern China. *J. Hydrol.* **2019**, *576*, 262–272. [CrossRef]

68. Wang, P.; Grinevsky, S.O.; Pozdniakov, S.P.; Yu, J.; Dautova, D.S.; Min, L.; Du, C.; Zhang, Y. Application of the water table fluctuation method for estimating evapotranspiration at two phreatophyte-dominated sites under hyper-arid environments. *J. Hydrol.* **2014**, *519 Pt B*, 2289–2300. [CrossRef]

69. Hill, M.C.; Tiedeman, C.R. *Effective Groundwater Model Calibration: With Analysis of Data, Sensitivities, Predictions and Uncertainty*; Wiley and Sons: Hoboken, NJ, USA, 2007.

70. Maddock Iii, T.; Baird, K.J.; Hanson, R.T.; Schmid, W.; Ajami, H. *RIP-ET: A Riparian Evapotranspiration Package for MODFLOW-2005*; 6-A39; US Geological Survey Techniques and Methods 6; U.S. Geological Survey: Reston, VA, USA, 2012; pp. 1–76.

71. Wang, P.; Niu, G.Y.; Fang, Y.H.; Wu, R.J.; Yu, J.J.; Yuan, G.F.; Pozdniakov, S.P.; Scott, R.L. Implementing Dynamic Root Optimization in Noah—MP for Simulating Phreatophytic Root Water Uptake. *Water Resour. Res.* **2018**, *54*, 1560–1575. [CrossRef]

Article

Coupling SWAT Model and CMB Method for Modeling of High-Permeability Bedrock Basins Receiving Interbasin Groundwater Flow

Javier Senent-Aparicio [1,*], Francisco J. Alcalá [2,3], Sitian Liu [1] and Patricia Jimeno-Sáez [1]

[1] Department of Civil Engineering, Universidad Católica San Antonio de Murcia, Campus de los Jerónimos s/n, Guadalupe, 30107 Murcia, Spain; sliu@alu.ucam.edu (S.L.); pjimeno@ucam.edu (P.J.-S.)

[2] Geological Survey of Spain, Ríos Rosas, 23 28003 Madrid, Spain; fj.alcala@igme.es

[3] Instituto de Ciencias Químicas Aplicadas, Facultad de Ingeniería, Universidad Autónoma de Chile, 7500138 Santiago, Chile

[*] Correspondence: jsenent@ucam.edu; Tel.: +34-968-278-818

Received: 10 February 2020; Accepted: 27 February 2020; Published: 29 February 2020

Abstract: This paper couples the Soil and Water Assessment Tool (SWAT) model and the chloride mass balance (CMB) method to improve the modeling of streamflow in high-permeability bedrock basins receiving interbasin groundwater flow (IGF). IGF refers to the naturally occurring groundwater flow beneath a topographic divide, which indicates that baseflow simulated by standard hydrological models may be substantially less than its actual magnitude. Identification and quantification of IGF is so difficult that most hydrological models use convenient simplifications to ignore it, leaving us with minimal knowledge of strategies to quantify it. The Castril River basin (CRB) was chosen to show this problematic and to propose the CMB method to assess the magnitude of the IGF contribution to baseflow. In this headwater area, which has null groundwater exploitation, the CMB method shows that yearly IGF hardly varies and represents about 51% of mean yearly baseflow. Based on this external IGF appraisal, simulated streamflow was corrected to obtain a reduction in the percent bias of the SWAT model, from 52.29 to 22.40. Corrected simulated streamflow was used during the SWAT model calibration and validation phases. The Nash–Sutcliffe Efficiency (NSE) coefficient and the logarithmic values of NSE (lnNSE) were used for overall SWAT model performance. For calibration and validation, monthly NSE was 0.77 and 0.80, respectively, whereas daily lnNSE was 0.81 and 0.64, respectively. This methodological framework, which includes initial system conceptualization and a new formulation, provides a reproducible way to deal with similar basins, the baseflow component of which is strongly determined by IGF.

Keywords: SWAT model; CMB method; interbasin groundwater flow; Castril River; baseflow filter

---

1. Introduction

Hydrological models have become essential tools for water management issues due to their ability to simulate the hydrological cycle through integrated and multidisciplinary approaches, along with their skills to simulate climate change scenarios, land use, and water management [1]. Reliability of such models depends on the spatial and temporal scales covered, as well as the capacity to conceptualize the system functioning [2–4]. Those hydrological models operating at a basin scale are powerful decision support tools because they can provide insights into water resource management [5]. Among them, the SWAT (Soil and Water Assessment Tool) model, a physically-based and semi-distributed eco-hydrological open access model [6] stands out. SWAT can simulate the quality and quantity of surface water and groundwater balance components at different catchment scales to predict the impact of climate change on the water balance of large watersheds [7,8] and deduce the effect of human-induced actions on water resources, such as irrigation practices and land-use changes [9]. A recent review of water quality and erosion models reveals that SWAT is, by far, the most used model [10]. However, a downside of SWAT is related to the simplified groundwater concept [11]. The simplified representation of the groundwater discharge and aquifer storage processes has been highlighted by several authors as something that may lead to a misunderstanding of the hydrological processes that occur in groundwater dominated watersheds [12].

One of these limitations is SWAT's inability to consider interbasin groundwater flow (IGF). IGF can be defined as the naturally occurring groundwater flow beneath the topographic divide that defines the basin boundary introduced in the SWAT model and in other hydrological models. It contributes to the baseflow of another basin different from that from which it was generated [13]. The magnitude of IGF may be especially relevant in high-permeability bedrock areas, such as steep karst areas. IGF maintains permanent streamflow in dry seasons, thus significantly altering the water balance of a region [14]. Despite the fact that IGF is a common hydrological process in large karst areas, it is often difficult to estimate, even tentatively [15]. Methodologies to identify and quantify IGF have traditionally relied on physical techniques, including the soil–water budget and water fluctuation, when there are sufficient data [16,17], tracer techniques measuring mostly environmental chemicals and stable isotope contents of precipitation and stream water [18,19], and groundwater modeling tools for indirect evaluations [20,21]. Some studies have aimed to assess IGF using the SWAT model. More specifically, Palanisamy and Workman (2014) [22] developed the KarstSWAT model to simulate IGF in watersheds dominated by typical karst features, determining input (recharge in sinkholes) and output (discharge in springs) water component dynamics. Malagó et al. (2016) [23] developed the KSWAT model, which was based on a combination of two previous SWAT applications: (1) a SWAT model adaptation to consider fast infiltration through caves and sinkholes up to the deep aquifers developed by Baffaut and Benson (2009) [24] and (2) a karst flow model in Excel to simulate spring flow discharge developed by Nikolaidis et al. (2013) [25]. More recently, Nguyen et al. (2020) [14] proposed a two linear reservoir model to represent the duality of aquifer recharge and discharge processes in a karst-dominated area in Germany. However, this interest is ongoing and, to our knowledge, the SWAT model has yet to be combined with the CMB method to improve hydrological cycle simulation in those basins where there is a difference between groundwater flow divides and surface topographic divides.

The evaluation of IGF is a complex, uncertain task when groundwater system functioning is partially unknown and the spatiotemporal coverage of data is too low to implement suitable evaluation techniques. In general, spatiotemporal coverage of environmental variables decreases in mountainous areas, thereby limiting the range of suitable techniques to assess IGF and other water balance components. In ungauged areas, IGF can be indirectly assessed when enough is known about groundwater system functioning to assert that IGF equals net aquifer recharge, which is the typical circumstance in most mountainous karst areas in a natural regime. In steep basins with gaining streams under a natural (undisturbed) regime, long-term net aquifer recharge (R) and discharge can be equated when groundwater abstraction, direct evapotranspiration from shallow aquifers, and underflow to deep aquifers are virtually null [26,27]. In such undisturbed hydrological functioning, net aquifer discharge equals the baseflow component of streamflow [28–31], and the problem shrinks to a matter of implementing suitable and viable techniques to determine R. Note that R is the infiltration amount that effectively contributes to the aquifer storage after some delay, smoothing out the variability inherent to precipitation events [32,33]. To assess renewable groundwater resources that finally reach streambeds, R is the governing variable [4,34].

Different methods can be used for R [35,36]. An independent, well-known method to determine R is the atmospheric Chloride Mass Balance (CMB) method [37–41]. The CMB method has been widely applied in different orographic, climatic, and geological contexts to yield mostly long-term (steady) R estimates when recharge water salinity can be attributed to the atmospheric salinity that reaches the water table. The CMB method was recently used to assess distributed mean R from precipitation and its uncertainty over continental Spain by verifying that the CMB variables were steady long-term [34,42].

This data availability was the reason the CMB method was chosen to assess IGF in areas with no human activities. In other territories, other techniques strictly intended for regional R can be selected for IGF evaluations when there are available data sets of similarly sufficient confidence.

This paper aims to evaluate the reliability of coupling the SWAT model and the CMB method to improve the modeling of streamflow in high-permeability bedrock basins receiving IGF. To that end, two main methodological steps are introduced. The first conceptualizes the hydrogeological functioning to confidently estimate IGF from existing CMB datasets for a control period and introduces a new formulation to generate a long-term baseflow series. The second step integrates corrected baseflow series into the SWAT model to improve the streamflow simulation. This methodology has been applied to the Castril River basin (CRB), which is an undisturbed, high-permeability bedrock area, characterized by the strong contribution of IGF to streamflow, as evidenced by a preliminary surface runoff coefficient greater than one.

2. Materials and Methods

2.1. Study Area

The Castril River is an aquifer-fed mountain stream located 37°47′–37°59′ north and 2°40′–2°50′ west at the headwater of the Guadalquivir River watershed (GRW) in the province of Granada in southern Spain, adjacent to the Segura River watershed (SRW) (Figure 1a). The Castril River basin (CRB) headwater extends from the GRW-SRW divide (peak elevation is 2130 m a.s.l. in the north) to the Portillo Reservoir (outlet is 837 m a.s.l. in the south), covers an area of about 120 km^2, and flows southward among the Sierra de Castril (west) and Sierra Seca (east) Mountains [43] (Figure 1b).

The climate is comparable to the continental Mediterranean, according to the Köppen classification [44]. Average annual precipitation is about 770 mm, with a coefficient of variation of 0.31 over the period 1951–2017. Precipitation is generated by Atlantic weather fronts coming in from the west and by short, intense Mediterranean convective storms. Most precipitation occurs during the autumn and spring. In winter, wet westerly and cold northerly winds predominate, whilst in summer and autumn, wet easterly and warm southerly winds blow [45]. Based on the period 1951–2016, the average annual temperature is about 8 °C, with the lowest temperatures in January and the highest in August. Average annual potential evapotranspiration is about 800 mm [46].

Geologically, the area belongs to the inner Prebetic domain of the external zone of the Alpine Betic Chain, which includes the following synthetic succession from bottom to top [47,48]: (1) Triassic gypsum-rich marls and clays (Keuper facies) with occasional limestones (Muschelkalk facies); (2) Lower and Middle Jurassic dolomites and oolitic limestones; (3) Upper Jurassic nodule limestones, calcareous marls, and marls; (4) Lower Cretaceous dolomites, dolomitic limestones, and marls; (5) Upper Cretaceous calcareous marls, marls, and dolomitic limestones; and (6) Paleocene to Middle Miocene limestones, marls, and calcarenites [48,49]. Small Late Quaternary alluvial deposits intermittently fill the valleys (Figure 1c).

From a hydrogeological point of view, geological materials can be classified into four groups, attending to the permeability type and storage capacity reported by the Geological Survey of Spain (IGME) (1988, 1995, 2001) [50–52]: (1) Triassic marls and clays are low-permeability materials that form the impervious boundary of local aquifers; (2) Jurassic and Cretaceous carbonate materials form highly permeable aquifers as thick as 300 m and have manifest karst features; (3) Jurassic and Cretaceous marls and calcareous marls are low-permeability materials, often confining the above Jurassic and Cretaceous carbonate materials; and (4) Late Quaternary alluvial deposits form temporary unconfined aquifers (Figure 1c).

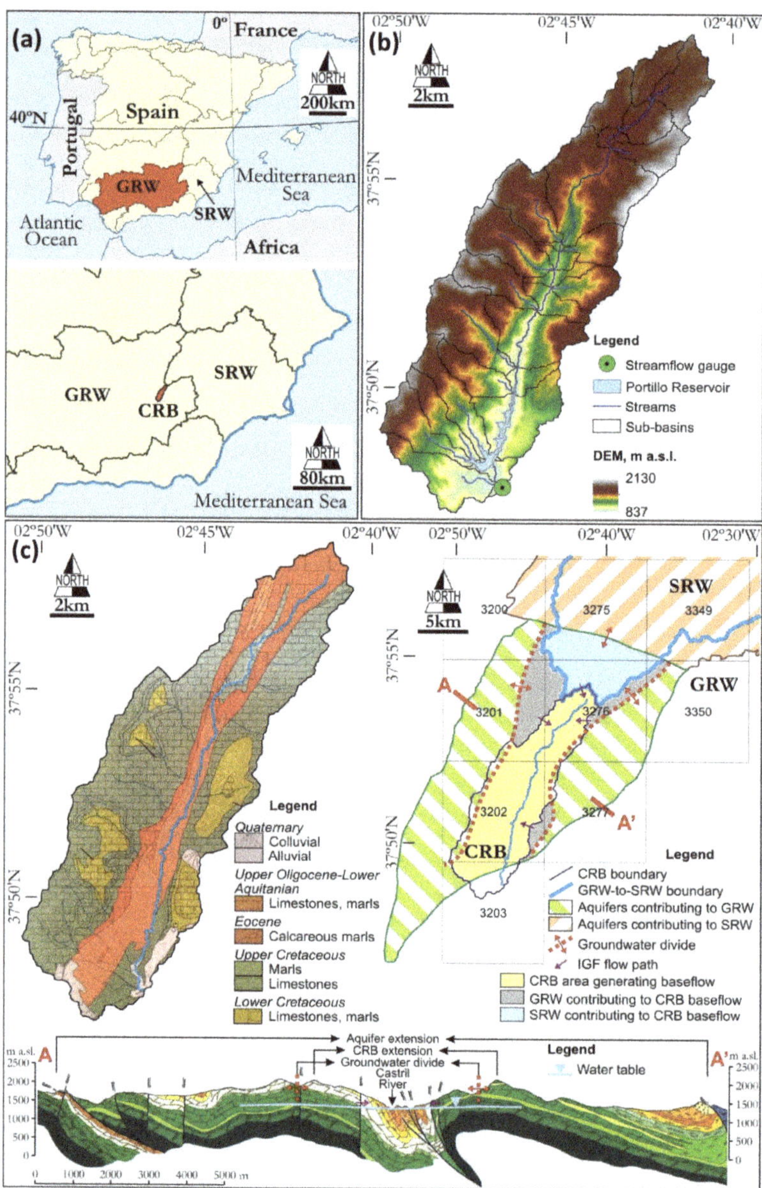

Figure 1. (**a**) Location of the Castril River basin (CRB) within the Guadalquivir River watershed (GRW) in southern Spain, adjacent to the Segura River watershed (SRW). (**b**) Discretization of the CRB and 29 sub-basins using the 25 m resolution Digital Elevation Model (DEM) from the Spanish National Geographic Institute, showing other features cited in the text. (**c**) After the Geological Survey of Spain (IGME) (1988, 1995, 2001) [50–52] and direct field observations, a hydrogeological map of the CRB (scale 1:200,000), the schematic hydrogeological functioning of the CRB and the hydraulically connected upstream GRW and SRW contributing areas, a hydrogeological cross-section A–A′ showing aquifer dimensions, CBR location, groundwater divides and flow paths, and the 10 km × 10 km cells for distributed net aquifer recharge (R) in the part of continental Spain [34,42] covered in the study area was developed.

Hydrogeological functioning of the area was defined by IGME (1988, 1995, 2001) [50–52]. Aquifer boundaries, groundwater divides, and groundwater flow paths were established from hydrogeological maps, piezometry, and chemical and isotopic data. These hydrogeological criteria enabled experts to identify preferred areas for aquifer recharge in summits and for aquifer discharge, at the precise place where the incisive valley topography intersects the piezometry of Cretaceous carbonate aquifers to generate intermittent (upstream) and permanent (downstream) springs (Figure 1c). Downstream, outside the study area, Pliocene and Quaternary alluvial formations fill the Castril River valley and form an unconfined aquifer that hydraulically connects to the stream [43].

The study area is within the Sierra de Castril Natural Park, which has been an environmentally protected space since 1989, and is catalogued as a zone of special conservation for wildlife by the European Natura 2020 network. With respect to land use, forest, grassland, woodland and shrubland, sparsely vegetated areas, bare areas, and marginal rain-fed crops occupy most of the basin surface (Figure 2). Other marginal uses are seasonal livestock (sheep and goats), irrigated traditional crops, and riverine forest of *Pinus nigra* and *Pinus Halepensis* [53]. Neither permanent human settlements nor relevant water uses exist.

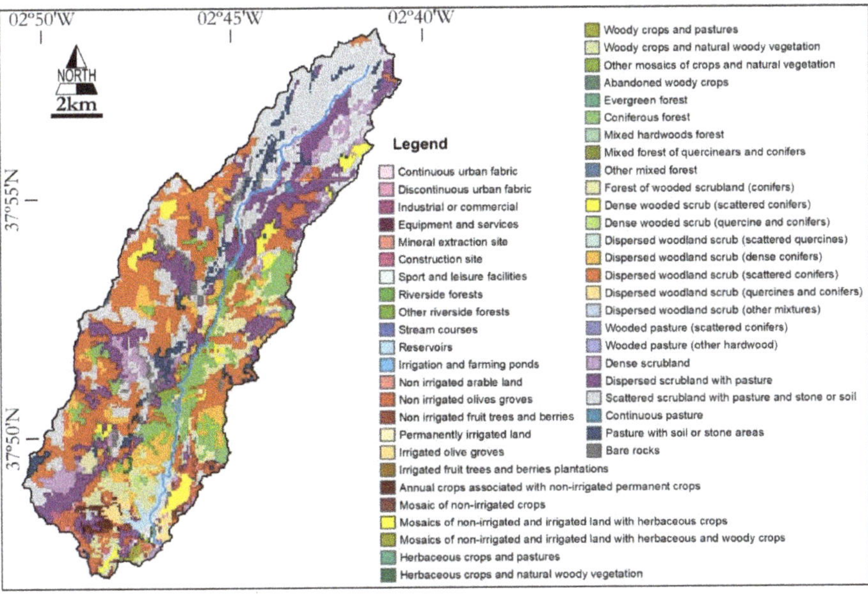

Figure 2. Land-use map (scale 1:25,000) from the Andalusian Environmental Information Network (REDIAM).

2.2. Overall Model Description

A coupled application of the SWAT model and the CMB method to improve streamflow simulation by considering IGF is introduced. This application includes four steps, shown as bulleted lists (Figure 3). The first step uses the CMB method to assess the magnitude of IGF contributing to the CRB baseflow from another upstream GRW and SRW areas, as described in Section 2.3. The second step uses the automated digital filter program BFLOW to split daily streamflow records into baseflow and surface runoff components, as a prerequisite to correct streamflow records by adding IGF to the baseline component, as described in Section 2.4. The third step uses the SWAT model to compare simulated streamflow with and without IGF, as described in Section 2.5. The fourth step uses the SWAT model for standard calibration and validation of simulated streamflow by considering the IGF correction, as described in Section 2.5.

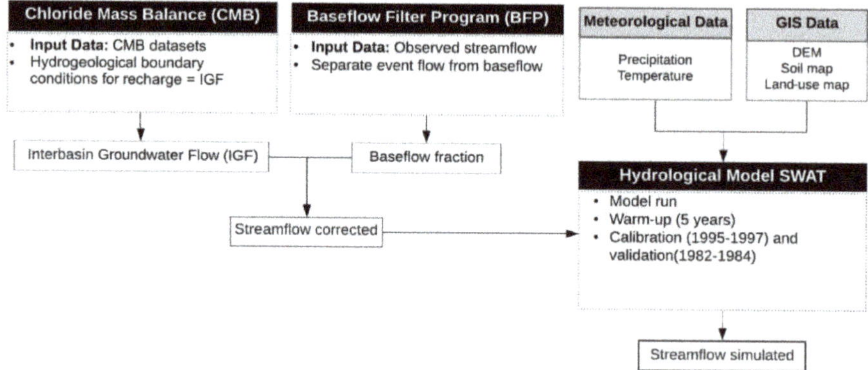

Figure 3. Flow diagram for the coupled Soil and Water Assessment Tool (SWAT) model and chloride mass balance (CMB) method application to model streamflow of hydrological basins subjected to interbasin groundwater flow (IGF).

2.3. CMB Method

2.3.1. CMB Method Application for Aquifer Recharge over Continental Spain

The atmospheric chloride mass balance (CMB) is one of the most widely used methods to estimate net aquifer recharge (R) from precipitation in different orographic, climatic, and geological contexts [35,36]. The CMB is a global method based on the principle of mass conservation of a conservative tracer, in this case the chloride ion, atmospherically contributing to the land surface. This technique yields mostly long-term (steady) R estimations when recharge water salinity can be attributed to the atmospheric salinity that reaches the water table [37–41].

The CMB method was recently applied to estimate distributed spatial mean R and its natural uncertainty (standard deviation) over continental Spain. For a confident application, the long-term steady condition of the CMB variables: atmospheric chloride bulk deposition, chloride export flux by surface runoff, and recharge water chloride content was verified [34,42,54]. This evaluation examined the influence of hydraulic properties (mostly permeability and storability) of different aquifer lithologies on R estimates, as well as the potential contribution of non-atmospheric sources of chloride [55]. For local usage, the reliability and hydrological meaning of distributed R were evaluated by comparing them with local, presumably trustworthy R estimates; one of these local cases was the CRB. Ordinary kriging was used to regionalize the CMB variables at the same 4976 nodes of a 10 km × 10 km grid. In each grid node a mean R value was estimated. Nodal R values were affected by two main types of uncertainty, the natural variability of the CMB variables and the error from its mapping. These uncertainties were identified and estimated [34,42].

The evaluation covered a 10-year period, which represented the critical balance period for the CMB variables to reach comparable steady means and standard deviations. This 10-year period matches the decadal global climatic cycles acting on the Iberian Peninsula, with irregular ~5-year positive and negative phases that follow the North Atlantic Oscillation trend [45,56]. Considering that (1) at least a 10-year balance period is required for reliably steady R evaluations in continental Spain and (2) the CMB datasets preferably spanned the period 1994–2007, the control period (1996–2005), which span a full 10-year long NAO climatic cycle, was chosen to estimate R in this work. Other authors have also implemented these CMB datasets for reliable local R evaluations in different climatic and geological settings, such as Andreu et al. (2011) [27] in Sierra de Gádor karst Massif in Southern Spain, Raposo et al. (2013) [57] in varied geological contexts in Galicia, on the coast of Northern Spain, and Barberá et al. (2018) [58] in the high-mountain, weathered-bedrock Bérchules basin in Southern Spain.

2.3.2. IGF Series Generation

As introduced in Section 1, long-term steady R and IGF can be made equivalent in steep basins with gaining stream under natural (undisturbed) regime when groundwater abstractions, direct evapotranspiration from shallow aquifers, and underflow to deep aquifers are virtually null, and the hydrogeological functioning is well-defined [28–31]. This is the case of the CRB, as described in Section 2.1, because human water use is virtually null and there is enough hydrogeological information. The fraction of R produced in upstream contributing areas can be used as a reliable proxy for the additional baseflow fraction contributing to the CRB baseline [26,27]. To assess the IGF, R is the significant factor [4,34].

As described above, several cells, each one yielding a nodal mean R value and its standard deviation for the control period (1996–2005) instead of a yearly R series, cover the CRB and upstream contributing areas. Adapted from Pulido-Velázquez et al. (2018) [59], a procedure is introduced to obtain the yearly R series by adopting the temporal structure of the yearly P series for the control period (1996–2005). This model uses a correction function that forces the control R series to have the same relative deviation as the control P series, while maintaining the magnitude of its initial mean and standard deviation. The calibrated function is applied to obtain a yearly R series, presuming the correction function does not change. The calculation includes the following steps:

Average change in mean and standard deviation of P and R for the same control period (1996–2005):

$$\Delta m = \frac{m(R) - m(P)}{m(P)} \tag{1a}$$

$$\Delta \sigma = \frac{\sigma(R) - \sigma(P)}{\sigma(P)} \tag{1b}$$

where Δm is the change in mean and $\Delta \sigma$ is the change in standard deviation.

Normalization of the yearly P series:

$$Pn_i = \frac{P_i - \overline{P}}{\sigma_R} \tag{2}$$

where P_i is i–year P and Ps_i is its normalized value, $\overline{P}$ is mean P, and σ_R is standard deviation of mean R.

Generation of yearly R series from yearly P series:

$$R_i = m_C + \sigma_C \cdot Ps_i \tag{3}$$

where R_i is i–year R, and m_C and σ_C are expressed as:

$$m_C = m(P) \cdot (1 + \Delta m) \tag{4a}$$

$$\sigma_C = \sigma(P) \cdot (1 + \Delta \sigma) \tag{4b}$$

When Equation (4) is applied to the control yearly P series, the generated yearly R series adopts the same mean and standard deviation as the control R series from Alcalá and Custodio (2014, 2015) [34,42]. When this procedure is applied to the full yearly P series, the equivalent yearly R series is obtained by assuming that the bias correction remains invariant over the full observation period.

2.4. BFLOW Program

The automated digital filter program (BFLOW) to split daily streamflow records into the baseflow and surface runoff components was used. Nathan and McMahon (1990) [60] were the first to implement this recursive digital filter technique for baseflow analysis. The hypothesis of BFLOW is that low-frequency signals represent the baseflow component while high-frequency signals represent the runoff component [61]. This technique gives results similar to those obtained using other automated models or manual techniques despite having no physical basis. BFLOW has been used in many studies related to the SWAT model [30,62]; see Arnold and Allen (1999) [63] for more details about this technique. The baseflow obtained by BFLOW was increased by adding IGF, calculated in previous section.

2.5. SWAT Model

2.5.1. Description of the SWAT Model

The Soil and Water Assessment Tool (SWAT) is a long-term watershed hydrological model with strong physical mechanisms, developed jointly in 1994 by the Agricultural Research Service of the United States Department of Agriculture (USDA-ARS) and Texas A&M AgriLife Research, part of the Texas A&M University System. The SWAT model simulation includes atmospheric precipitation, surface runoff, subsurface flow, evapotranspiration, groundwater flow, river network flow concentration, and other intermediate water balance components subjected to variable delays [6]. The SWAT model first divides the study area into several hydrological response units (HRUs) based on the digital elevation model (DEM), land use, soil type, and meteorological data. Then, SWAT establishes a hydrophysical conceptual model of each HRU, calculates runoff in each HRU, and finally connects the entire set of the HRU runoff responses through the river network of the study area toward the basin outlet. The hydrological processes simulated by the SWAT model are based on the following water balance equation:

$$SW_t = SW_0 + \sum_{i=1}^{t}\left(P_{day} - E_{day} - Q_{surf} - W_{seep} - Q_{gw}\right)_i \tag{5}$$

where SW_t is final soil–water content (mm H_2O), SW_0 is initial soil–water content (mm H_2O), t is time (days), P_{day} is precipitation on day i (mm H_2O), ET_{day} is evapotranspiration on day i (mm H_2O), Q_{surf} is surface runoff on day i (mm H_2O), W_{seep} is water amount that enters the vadose zone from the soil profile on day i (mm H_2O), and Q_{gw} is groundwater return flow on day i (mm H_2O).

2.5.2. Data, Model Set-Up, Calibration, and Validation

Datasets used to implement the SWAT model were: (1) the 25-m resolution DEM from the Spanish National Geographic Institute (IGN); (2) the land-use map (scale 1:25,000) from the Andalusian Environmental Information Network (REDIAM); (3) the 1-km resolution georeferenced soil data from the World Soil Coordination Map; (4) the 5-km resolution nodal daily precipitation series in Spain from the Spanish National Weather Service (AEMET) grid version 1.0, which cover the period 1951–2017; (5) the 10-km resolution nodal daily temperature series in Spain from the fifth version of the high-resolution SPAIN02 grid, which cover the period 1951–2016; and (6) the 24-h streamflow records downloaded from the Spanish Centre for Public Works Studies and Experimentation (CEDEX) website. The open source QGIS interface for SWAT (QSWAT 1.8) was used to set up the SWAT model.

The SUFI-2 algorithm of SWAT-CUP (Calibration and Uncertainty Programs) to calibrate and validate the SWAT model was used. Based on our previous modeling experiences [64,65], twenty-one widely used flow calibration parameters and their ranges were initially selected. Aimed at reaching an acceptable calibration, two iterations (representing 500 simulations each) were performed; the first included 13 parameters on a monthly scale, the latter included 8 parameters on a daily scale. To mitigate the effect of initial soil–water condition, a five-year warm-up period was imposed [66]. The periods 1995–1997 and 1982–1984 were, respectively, selected for the calibration and validation phases. As the downloaded daily streamflow (discharge) series was discontinuous, time intervals for calibration and validation were carefully selected to minimize the effect of existing data gaps.

As the CRB is a singular aquifer-fed mountain stream, some quantitative information to cross-validate the SWAT model results were used. For model efficiency criteria, Nash-Sutcliffe efficiency coefficient (NSE), logarithmic form of the NSE (lnNSE), coefficient of determination (R^2), percent bias (PBIAS), Root Mean Square Error (RMSE), and RMSE relative to standard deviation of the observed data (RSR) were used (Table 1).

Table 1. Equations, ranges, and optimal values for SWAT model performance statistics, after Moriasi et al. (2012) [67].

Statistic and Equation [1]	Description
NSE : Nash–Sutcliffe Efficiency Coefficient $$= 1 - \frac{\sum_{i=1}^{n}(Q_{obs\ i} - Q_{sim\ i})^2}{\sum_{i=1}^{n}(Q_{obs\ i} - \overline{Q})^2}$$	NSE indicates a perfect match between observed and simulated data, and ranges from $-\infty$ to 1. Higher than 0.5 is considered satisfactory.
$$\text{lnNSE} = 1 - \frac{\sum_{i=1}^{n}(\ln(Q_{obs\ i}) - ln(Q_{sim\ i}))^2}{\sum_{i=1}^{n}\left(\ln(Q_{obs\ i}) - \overline{\ln(Q)}\right)^2}$$	lnNSE is the logarithmic form of the model efficiency coefficient. NSE emphasizes the high flows, and lnNSE emphasizes the low flows.
R^2 : *Coefficient of Determination* $$= \left(\frac{\sum_{i=1}^{n}(Q_{obs\ i} - \overline{Q})(Q_{sim\ i} - \overline{Q_{sim\ i}})}{\sqrt{\sum_{i=1}^{n}(Q_{obs\ i} - \overline{Q})^2}\sqrt{\sum_{i=1}^{n}(Q_{sim\ i} - \overline{Q_{sim\ i}})^2}}\right)^2$$	R^2 indicates the degree of linear relationship between simulated and observed data, and ranges from 0 to 1. Higher than 0.5 is considered a satisfactory result.
PBIAS : Percent Bias $$= \frac{\sum_{i=1}^{n}(Q_{obs\ i} - Q_{sim\ i})\Delta 100}{\sum_{i=1}^{n}(Q_{obs\ i})}$$	PBIAS calculates the average tendency of the simulated data to be higher or lower than their observed counterparts. The optimal value is 0, and an acceptable one is between ± 25.
RMSE : Root Mean Square Error $$= \sqrt{\sum_{i=1}^{n}(Q_{obs\ i} - Q_{sim\ i})^2}$$	RMSE = 0 indicates a perfect match between observed and simulated data. Increasing RMSE values indicate that matching is getting worse.
RSR : Root Mean Square Error relative to standard deviation of the observed data $$= \frac{RMSE}{STDEV_{obs}} = \frac{\sqrt{\sum_{i=1}^{n}(Q_{obs\ i} - Q_{sim\ i})^2}}{\sqrt{\sum_{i=1}^{n}(Q_{obs\ i} - \overline{Q_{sim\ i}})^2}}$$	RSR is RMSE relative to standard deviation of the observed data, and ranges from 0 to ∞. The lower the RSR, the lower the RMSE and the better the model performance. Lower than 0.7 is acceptable.

[1] n is the total number of observations, $Q_{obs\ i}$ and $Q_{sim\ i}$ are observed and simulated streamflow at observation i, $\overline{Q}$ is the mean of the observed data over the simulation period, and $\overline{Q_{sim\ i}}$ is the mean of the simulated data over the simulation period.

3. Results and Discussion

3.1. Using the CMB Datasets to Estimate IGF

As shown in Figure 1c, the entire hydrogeological system that contributes to streamflow at the CRB outlet covers the CRB surface itself and some hydraulically connected adjacent areas from GRW and SRW. The methodology described in Section 2.3 was applied to the nodal R values gathered from those 10 km × 10 km cells covering the CRB and those contributing to upstream GRW and SRW areas. Attending to the hydrogeological functioning (Figure 1c) and existing land and water uses (Figure 2), nodal mean values and standard deviations of R and baseflow can be assumed to be equal.

In this area, for the control period (1996–2005), nodal mean R varied within the range 143–332 mm year^{-1}, which means recharge–precipitation ratios were in the 0.29–0.37 range; the standard deviation of mean R varied within the 39–90 mm year^{-1} range, which placed the given coefficients of variation of mean annual R (mean value-to-standard deviation ratio) in the 0.27–0.30 range (Table 2). For the control period (1996–2005), fitting parameters were calculated to generate the yearly R data series in the CRB and upstream GRW and SRW contributing areas, which are in Table 3, whereas the generated surface-weighted yearly P and R series are in Table 4. In each area, yearly R and P series for the control period (1996–2005) were compared. The resulting parametric functions allowed for the extension of the calculated yearly R series to cover the yearly P full record (1951–2016) (Figure 4). Figure 5 shows the full yearly baseflow series generated within the CBR, as well as the yearly surface-weighted IGF series contributed by upstream GRW and SRW areas. As observed, IGF is somewhat higher than baseflow, generated within the CRB. IGF is about 51% of total CRB baseflow.

Table 2. For the 10 km × 10 km cells covering the CRB and upstream GRW and SGW contributing areas, the CMB datasets gathered from Alcalá and Custodio (2014, 2015) [34,42].

Cell [1]	CRB					GRW					SRW				
	S	P [2]	CVP	R	CVR	S	P	CVP	R	CVR	S	P	CVP	R	CVR
3200						3.2	894	0.31	315	0.27	0.8	894	0.31	315	0.27
3201	10.4	909	0.33	332	0.27	20.2	909	0.33	332	0.27					
3202	60.6	693	0.34	229	0.28	4.8	693	0.34	229	0.28					
3203	7.1	486	0.35	143	0.27										
3275						1.0	813	0.32	276	0.27	21.7	813	0.32	276	0.27
3276	22.0	668	0.33	206	0.29	14.5	668	0.33	206	0.29	25.0	668	0.33	206	0.29
3277	1.8	517	0.35	153	0.30	1.1	517	0.35	153	0.30					
3349						0.1	687	0.32	212	0.27	0.5	687	0.32	212	0.27
3350						2.0	612	0.33	186	0.28	0.5	612	0.33	186	0.28
Sum	101.9					46.9					48.4				
SWA [3]		692	0.34	227	0.28		787	0.33	269	0.28		736	0.32	239	0.28

[1] Cell ID as in Figure 1c. [2] S is surface in km^2, P and R are, respectively, mean precipitation and mean net aquifer recharge over the control period (1996–2005) in mm year^{-1}; and CVP and CVR are the dimensionless coefficients of variation of mean P and R over the control period (1996–2005) as fractions. [3] SWA is surface-weighted average.

Table 3. Fitting parameters for the CRB and upstream GRW and SGW areas.

Parameter [1]	CRB	GRW	SRW
Δm	−0.67	−0.66	−0.67
$\Delta\sigma$	−0.73	−0.71	−0.72
m_C	227	269	239
σ_C	63.1	74.9	66.8

[1] Δm and $\Delta\sigma$ are dimensionless, and m_C and σ_C are in mm year^{-1}.

Table 4. For the control period (1996–2005), surface-weighted yearly series of (i) P and R in the CRB and in upstream GRW and SRW areas, and (ii) IGF from the GRW and SRW area contributing to CRB.

Year	P [1]	Psi [1]	R, CRB [1]	R, GRW	R, SRW	IGF, GRW+SRW [2]
1996	1037.9	1.76	338.5	401.4	357.1	378.9
1997	978.9	1.47	319.8	379.2	337.3	357.9
1998	472.3	−1.08	159.2	188.7	167.4	177.9
1999	575.4	−0.56	191.9	227.5	202.0	214.6
2000	669.5	−0.09	221.7	262.9	233.6	248.0
2001	742.6	0.28	244.9	290.4	258.1	274.0
2002	616.8	−0.35	205.0	243.1	215.9	229.3
2003	723.7	0.19	238.9	283.3	251.8	267.3
2004	641.5	−0.23	212.9	252.4	224.2	238.1
2005	406.9	−1.40	138.5	164.1	145.5	154.7
Mean [3]	686.6		227.1	269.3	239.3	240.1
SD	199.2		63.1	74.9	66.8	66.8
CV	0.29		0.28	0.28	0.28	0.28

[1] P and R are, respectively, annual precipitation and net aquifer recharge in mm year^{-1}, and Psi is dimensionless normalized yearly P. [2] IGF is interbasin groundwater flow in mm year^{-1}. [3] Mean and SD are mean and standard deviation over the control period (1996–2005) in mm year^{-1}, and CV is dimensionless coefficient of variation as a fraction.

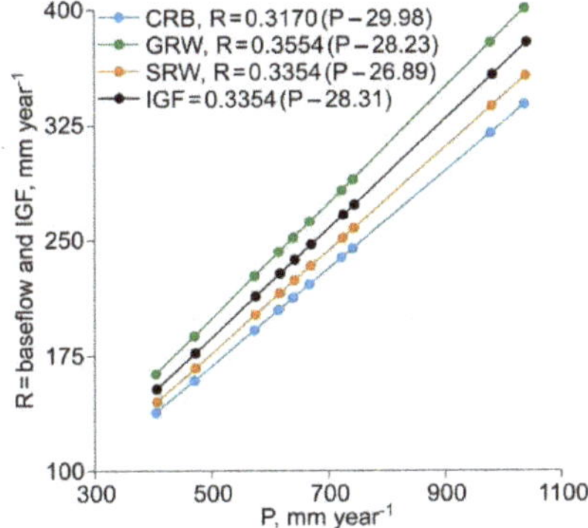

Figure 4. For the control period (1996–2005), parameterization of yearly P–R functions in the CRB and upstream GRW and SRW contributing areas; yearly R equals yearly baseflow. Yearly IGF series refers to the surface-weighed sum of upstream R = baseflow from GRW and SRW areas contributing to the CRB streamflow. In all cases, the Pearson coefficient of correlation is 1.

3.2. Comparison of SWAT Model Results with and without IGF

Based on DEM analysis and after SWAT model implementation, the CRB was discretized into 29 sub-basins. Based on the combination of land uses, soil types, and slope ranges (<2%, 2%–8%, >8%), 149 HRUs were defined. The thresholds for defining HRUs were set to 5% to optimize model processing. The Hargreaves non-global method was used to simulate potential evapotranspiration [68]. As a result, only precipitation and temperature data to run the SWAT model were needed.

As described in Section 3.1, the IGF from upstream GRW and SRW areas greatly contributes to the Castril River streamflow. For the period 1995–1997, the SWAT model was doubly implemented on a monthly scale with and without IGF. The result was a large difference between observed and initial simulated streamflow when IGF was omitted (Figure 6). When IGF was included as an additional baseflow fraction, the difference between observed and corrected simulated streamflow clearly narrowed. This preliminary trial at model performance showed that the statistics NSE and PBIAS improve when IGF was included (Figure 6). Overall model performance increased about 80% in absolute terms.

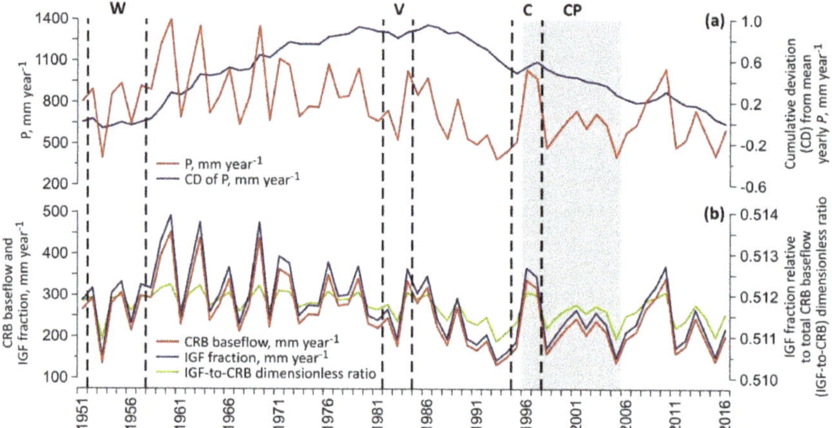

Figure 5. For the full period (1951–2016), (**a**) surface-weighted yearly P series in the area compiled from the Spanish National Weather Service (AEMET) grid version 1.0 and cumulative deviation (CD) from mean yearly P in mm year^{-1}; and (**b**) generated yearly baseflow series in the CRB and yearly surface-weighed IGF series from upstream GRW and SRW contributing areas in mm year^{-1}, and IGF fraction relative to total CRB baseflow (IGF–CRB) dimensionless ratio. The control period (1996–2005) is grey shallowed (CP). Vertical dotted lines indicate selected time intervals for the SWAT model warm-up (W), calibration (C), and validation (V) phases.

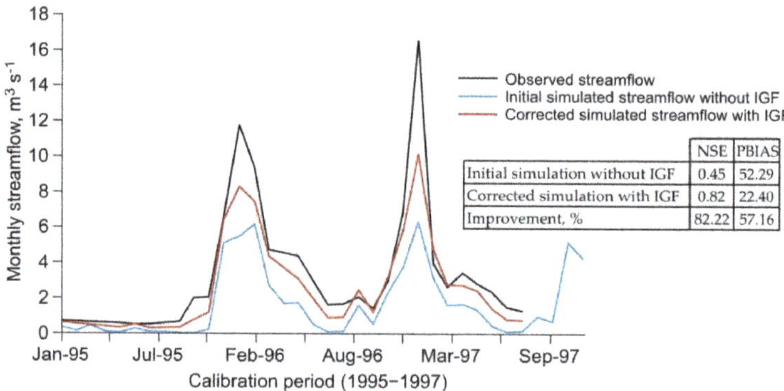

	NSE	PBIAS
Initial simulation without IGF	0.45	52.29
Corrected simulation with IGF	0.82	22.40
Improvement, %	82.22	57.16

Figure 6. For the selected calibration period (1995–1997) and on a monthly scale, observed streamflow compared to (i) initial simulated streamflow without IGF and (ii) corrected simulated streamflow with IGF. The statistics NSE and PBIAS show the model performance achieved in each simulation.

3.3. Calibration and Validation of SWAT Model Including IGF

A total of 21 SWAT parameters were optimized using the SUFI-2 algorithm from SWAT-CUP. As described in Section 2.5.2, parameter selection was based on our previous research experiences in similar basins in Southern Spain [64,65]. The SWAT calibration phase covered a 3-year period (1995–1997). The final ranges used and the final fitted values of these parameters are given in Table 5.

The magnitude of calibrated GW_REVAP, ESCO, LAT_TTIME, GWQMN, and ALPHA_BF parameters is quite similar to those obtained with similar orography, geology, climate, and land use [64,65]. The ESCO is also similar to that fitted in other Mediterranean karst areas, where yearly actual evapotranspiration is typically 0.7–0.9 fold yearly precipitation [26,27,46]. The low value of ALPHA_BF indicates a slow aquifer response [69]. This is corroborated by the long-delayed responses to recharge events in similar karst aquifers in the region, reported by Moral et al. (2008) [70].

Table 5. Description of parameters used for SWAT model calibration in the CRB.

Parameter [1]	Description	Range Used in Calibration	Fitted Value
r_CN2.mgt	Soil Conservation Service (SCS) runoff curve number	−0.1 to 0.1	0.08
v_ALPHA_BF.gw	Baseflow alpha factor (day^{-1})	0 to 1	0.11
a_GW_DELAY.gw	Groundwater delay time (day)	0 to 60	2.82
a_GWQMN.gw	Threshold depth of water in the shallow aquifer for return flow to occur (mm)	−200 to 1000	898.00
v_GW_REVAP.gw	Groundwater revap coefficient	0.02 to 0.1	0.09
a_RCHRG_DP.gw	Deep aquifer percolation fraction	−0.05 to 0.05	0.04
a_REVAPMN.gw	Threshold depth of water in shallow aquifer for revap or percolation to deep aquifer to occur (mm)	−500 to 500	−61.00
v_CANMX.hru	Maximum canopy storage (mm)	0 to 8	0.47
v_EPCO.bsn	Plant uptake compensation factor	0.5 to 1	0.56
v_ESCO.bsn	Soil evaporation compensation factor	0.3 to 0.8	0.61
r_SOL_AWC.sol	Available water capacity of the soil layer (mm H_2O/mm soil)	−0.02 to 0.02	−0.02
v_LAT_TTIME.hru	Lateral flow travel time (day)	0 to 180	76.50
v_SLSOIL.hru	Slope length for lateral subsurface flow (m)	0 to 150	1.35
r_SLSUBBSN.hru	Average slope length (m)	−0.5 to 0.5	0.08
r_HRU_SLP.hru	Average slope steepness (m/m)	−0.5 to 0.5	0.40
v_OV_N.hru	Manning's 'n' value for overland flow	0.01 to 1	0.61
r_CH_S1.sub	Average slope of tributary channels (m/m).	−0.5 to 0.5	0.26
v_CH_N1.sub	Manning's 'n' value for the tributary channels	0.01 to 30	1.68
r_CH_S2.rte	Average slope of main channel along the channel length (m/m)	−0.5 to 0.5	−0.04
v_CH_N2.rte	Manning's 'n' value for the main channel	0.01 to 0.3	0.04
v_SURLAG.bsn	Surface runoff lag coefficient	0.05 to 24	20.71

[1] (r_) refers to relative change, i.e., the current parameter must be multiplied by (1 + the value obtained in calibration), (v_) means that the existing parameter value must be replaced by the value obtained in calibration, and (a_) refers to absolute change, i.e., the fitted value must be added to the existing value of the parameter.

Corrected streamflow records with IGF were used for model calibration (1995–1997) and validation (1982–1984) phases. Observed streamflow was compared to corrected simulated streamflow on monthly (Figure 7) and daily (Figure 8) scales during the calibration and validation periods. In the CRB, the fitted SWAT model replicated, almost identically, the trend of the streamflow hydrograph. The higher fluctuations in the simulated peaks and the lower ones in low flows were found, both in monthly and daily streamflow simulations.

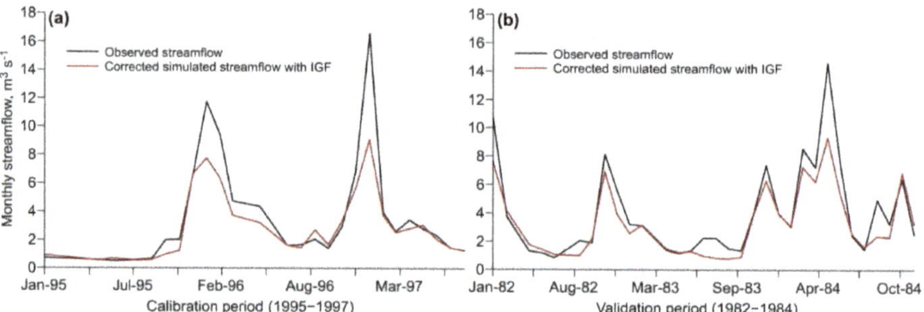

Figure 7. On a monthly scale, observed streamflow compared to corrected simulated streamflow with SWAT model for the (**a**) calibration and (**b**) validation phases.

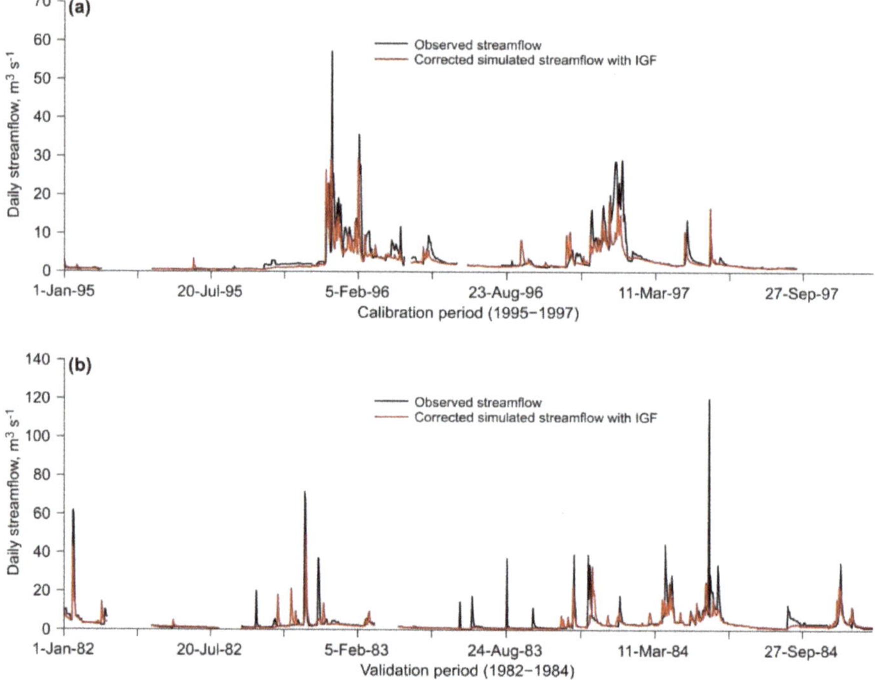

Figure 8. On a daily scale, observed streamflow compared to corrected simulated streamflow with SWAT model for the (**a**) calibration and (**b**) validation phases.

As observed in Figure 8, low flows predominate in the CRB daily streamflow record. As many other SWAT models reported for other similar aquifer-fed karst areas, the SWAT model performance for high and normal flows decreased in the face of predominant low flows [65,71]. Therefore, following the suggestion of Krause et al. (2005) [72], NSE and lnNSE were used to measure, respectively, the high and the low flows, to reduce the problem of the squared differences and the resulting sensitivity to extreme values of NSE. Calibration and validation of monthly corrected streamflow showed good agreement of simulated and observed data, as indicated by the model performance statistics for monthly and daily simulations given in Table 6.

Table 6. SWAT model performance statistics for corrected simulated monthly and daily streamflow during calibration and validation phases.

Statistic	Time Step	Calibration	Validation
NSE	Monthly	0.77	0.8
R^2	Monthly	0.92	0.89
PBIAS	Monthly	19.82	17.25
RSR	Monthly	0.48	0.44
lnNSE	Daily	0.81	0.64

As finally deduced, the SWAT model performs well and can be used for further analysis in the CRB and in other similar high-permeability bedrock basins, the baseflow of which is strongly determined by IGF. For this, there must be a confident evaluation of IGF or, at minimum, a reliable external evaluation available. In this paper, the CMB method and the available CMB datasets in continental Spain [34,42] were used for this purpose. However, as described for other groundwater and surface water coupling models, the SWAT-CMB application presented here reveals the following overall problems: (1) the spatial (size and volume) and temporal (renovation rate) scales of groundwater and surface water bodies differ, whereas the coupling model can only simulate the same spatial and temporal scale in both types of bodies; (2) the coupling models had defects in the coupling mechanism processing, which demanded substantial simplification of the coupling process, thereby causing model distortion; and (3) this coupling model was established for a certain region or a specific problem and, although good results have been achieved, there is no general adaptability, so additional hydrogeological knowledge of local applications is needed to consider the changes in scale effects and actual flow conditions.

4. Conclusions

This paper presents the combined application of the SWAT model and the CMB method to model streamflow in the CRB, a representative high-permeability bedrock basin where the streamflow is significantly determined by IGF from upstream contributing areas. The CMB method and available CMB datasets in continental Spain were used for the IGF that adds to the baseflow generated within the CRB. The SWAT model performance improved noticeably when simulated streamflow with IGF was used. Using the CMB datasets for streamflow correction, the SWAT model showed good performance both in daily and monthly simulations. Some overall remarks from this research are highlighted below.

The influence of IGF on basins like the CRB is remarkable. Therefore, IGF must be considered to improve water resource evaluation and management in this kind of basin located in the headwater of large river watersheds. In the CRB, IGF means about 51% of total baseflow. We do not suggest using the SWAT model alone for modeling of these aquifer-fed mountain basins. It must be coupled with other specific methods to accurately assess IGF. The CMB was revealed to be a suitable method for IGF, because of the favorable hydrogeological setting and the negligible groundwater abstraction, which allowed for equivalent net aquifer recharge to the baseflow contributing to streamflow in the headwater of large river watersheds. In other areas that reflect different patterns of groundwater use and hydrogeological features, assessment of IGF must rely on other techniques coupled with the SWAT model.

Author Contributions: J.S.-A. and F.J.A. conceived and designed the research; P.J.-S. and S.L. implemented the coupled application; S.L. compiled data. All authors contributed to writing the manuscript. All authors have read and agreed to the published version of the manuscript.

Funding: This research received no external funding.

Acknowledgments: The authors acknowledge the AEMET, IGN, IGME, CEDEX, and REDIAM Public Institutions for the data provided through their information service websites. We also acknowledge Papercheck Proofreading and Editing Services. We also wish to express our gratitude to two anonymous reviewers for their valuable advice and comments.

Conflicts of Interest: The authors declare no conflicts of interest.

References

1. Trolle, D.; Hamilton, D.P.; Hipsey, M.R.; Bolding, K.; Bruggeman, J.; Mooij, W.M.; Janse, J.H.; Nielsen, A.; Jeppesen, E.; Elliott, A.; et al. A community–based framework for aquatic ecosystem models. *Hydrobiologia* **2012**, *683*, 25–34. [CrossRef]

2. Hojberg, A.L.; Refsgaard, J.C. Model uncertainty-parameter uncertainty versus conceptual models. *Water Sci. Technol.* **2005**, *52*, 177–186. [CrossRef] [PubMed]

3. Beven, K. Towards integrated environmental models of everywhere: Uncertainty, data and modelling as a learning process. *Hydrol. Earth Syst. Sci.* **2007**, *11*, 460–467. [CrossRef]

4. Alcalá, F.J.; Martínez-Valderrama, J.; Robles-Marín, P.; Guerrera, F.; Martín-Martín, M.; Raffaelli, G.; Tejera de León, J.; Asebriy, L. A hydrological-economic model for sustainable groundwater use in sparse-data drylands: Application to the Amtoudi Oasis in southern Morocco, northern Sahara. *Sci. Total Environ.* **2015**, *537*, 309–322. [CrossRef]

5. Francesconi, W.; Srinivasan, R.; Pérez-Miñana, E.; Willcock, S.P.; Quintero, M. Using the Soil and Water Assessment Tool (SWAT) to model ecosystem services: A systematic review. *J. Hydrol.* **2016**, *535*, 625–636. [CrossRef]

6. Arnold, J.G.; Srinivasan, R.; Muttiah, R.S.; Williams, J.R. Large area hydrologic modeling and assessment Part I: Model development. *J. Am. Water Resour. Assoc.* **1998**, *34*, 73–89. [CrossRef]

7. Molina-Navarro, E.; Andersen, H.E.; Nielsen, A.; Thodsen, H.; Trolle, D. Quantifying the combined effects of land use and climate changes on stream flow and nutrient loads: A modelling approach in the Odense Fjord catchment (Denmark). *Sci. Total Environ.* **2018**, *621*, 253–264. [CrossRef]

8. Blanco-Gómez, P.; Jimeno-Sáez, P.; Senent-Aparicio, J.; Pérez-Sánchez, J. Impact of Climate Change on Water Balance Components and Droughts in the Guajoyo River Basin (El Salvador). *Water* **2019**, *11*, 2360. [CrossRef]

9. Senent-Aparicio, J.; Liu, S.; Pérez-Sánchez, J.; López-Ballesteros, A.; Jimeno-Sáez, P. Assessing Impacts of Climate Variability and Reforestation Activities on Water Resources in the Headwaters of the Segura River Basin (SE Spain). *Sustainability* **2018**, *10*, 3277. [CrossRef]

10. Fu, B.; Merritt, W.S.; Croke, B.F.W.; Weber, T.R.; Jakeman, A.J. A review of catchment-scale water quality and erosion models and a synthesis of future prospects. *Environ. Model. Softw.* **2019**, *114*, 75–97. [CrossRef]

11. Luo, Y.; Arnold, J.; Allen, P.; Chen, X. Baseflow simulation using SWAT model in an inland river basin in Tianshan Mountains, Northwest China. *Hydrol. Earth Syst. Sci.* **2012**, *16*, 1259–1267. [CrossRef]

12. Ficklin, D.L.; Luo, Y.; Zhang, M. Watershed Modelling of Hydrology and Water Quality in the Sacramento River Watershed, California. *Hydrol. Process.* **2012**, *27*, 236–250. [CrossRef]

13. Genereux, D.P.; Jordan, M.T.; Carbonell, D. A pired-watershed budget study to quantify interbasin groundwater flow in a lowland rain forest. Costa Rica. *Water Resour. Res.* **2005**, *41*, W04011. [CrossRef]

14. Nguyen, V.T.; Dietrich, J.; Uniyal, B. Modeling interbasin groundwater flow in karst areas: Model development, application, and calibration strategy. *Environ. Modell. Softw.* **2020**, *124*, 104606. [CrossRef]

15. Zanon, C.; Genereux, D.P.; Oberbauer, S.F. Use of a watershed hydrologic model to estimate interbasin groundwater flow in a Costa Rican rainforest. *Hydrol. Process.* **2014**, *28*, 3670–3680. [CrossRef]

16. Rahayuningtyas, C.; Wu, R.S.; Anwar, R.; Chiang, L.C. Improving avswat stream flow simulation by incorporating groundwater recharge prediction in the upstream Lesti watershed, East Java, Indonesia. *Terr. Atmos. Ocean. Sci.* **2014**, *25*, 881–892. [CrossRef]

17. Han, M.; Zhao, C.Y.; Šimůnek, J.; Feng, G. Evaluating the impact of groundwater on cotton growth and root zone water balance using Hydrus-1D coupled with a crop growth model. *Agric. Water Manag.* **2015**, *160*, 64–75. [CrossRef]

18. Obuobie, E. Estimation of Groundwater Recharge in the Context of Future Climate Change in the White Volta River Basin, West Africa. Germany. Ph.D. Thesis, Rheinischen Friedrich-Wilhelms-Universität, Bonn, Germany, 2008; 165p.

19. Alcalá, F.J.; Martín-Martín, M.; Guerrera, F.; Martínez-Valderrama, J.; Marín, P.R. A feasible methodology for groundwater resource modelling for sustainable use in sparse-data drylands: Application to the Amtoudi Oasis in the northern Sahara. *Sci. Total Environ.* **2018**, *630*, 1246–1257. [CrossRef]

20. Kim, N.W.; Chung, I.M.; Won, Y.S.; Arnold, J.G. Development and application of the integrated SWAT-MODFLOW model. *J. Hydrol.* **2008**, *356*, 1–16. [CrossRef]

21. Bouaziz, L.; Weerts, A.; Schellekens, J.; Sprokkereef, E.; Stam, J.; Savenije, H.; Hrachowitz, M. Redressing the balance: Quantifying net intercatchment groundwater flows. *Hydrol. Earth Syst. Sci.* **2018**, *22*, 6415–6434. [CrossRef]

22. Palanisamy, B.; Workman, S.R. Hydrologic modeling of flow through sinkholes located in streambeds of Cane Run stream, Kentucky. *J. Hydrol. Eng.* **2014**, *20*, 04014066. [CrossRef]

23. Malagó, A.; Efstathiou, D.; Bouraoui, F.; Nikolaidis, N.P.; Franchini, M.; Bidoglio, G.; Kritsotakis, M. Regional scale hydrologic modeling of a karst-dominant geomorphology: The case study of the island of Crete. *J. Hydrol.* **2016**, *540*, 64–81. [CrossRef]

24. Baffaut, C.; Benson, V.W. Modeling flow and pollutant transport in a karst watershed with SWAT. *Trans. ASABE* **2009**, *52*, 469–479. [CrossRef]

25. Nikolaidis, N.P.; Bouraoui, F.; Bidoglio, G. Hydrologic and geochemical modeling of a karstic Mediterranean watershed. *J. Hydrol.* **2013**, *477*, 129–138. [CrossRef]

26. Alcalá, F.J.; Cantón, Y.; Contreras, S.; Were, A.; Serrano-Ortiz, P.; Puigdefábregas, J.; Solé-Benet, A.; Custodio, E.; Domingo, F. Diffuse and concentrated recharge evaluation using physical and tracer techniques: Results from a semiarid carbonate massif aquifer in southeastern Spain. *Environ. Earth Sci.* **2011**, *63*, 541–557. [CrossRef]

27. Andreu, J.M.; Alcalá, F.J.; Vallejos, Á.; Pulido-Bosch, A. Recharge to aquifers in SE Spain: Different approaches and new challenges. *J. Arid Environ.* **2011**, *75*, 1262–1270. [CrossRef]

28. Rutledge, A.T.; Mesko, T.O. *Estimated Hydrologic Characteristics of Shallow Aquifer Systems in the Valley and Ridge, the Blue Ridge, and the Piedmont Physiographic Provinces Based on Analysis of Streamflow Recession and Base Flow*; U.S. Geological Survey: Reston, VA, USA, 1996; 58p.

29. Lim, K.J.; Engel, B.A.; Tang, Z.; Choi, J.; Kim, K.S.; Muthukrishnan, S.; Tripathy, D. Automated web GIS based hydrograph analysis tool, WHAT. *J. Am. Water Resour. Assoc.* **2005**, 1407–1416. [CrossRef]

30. Plesca, I.; Timbe, E.; Exbrayat, J.-F.; Windhorst, D.; Kraft, P.; Crespo, P.; Vaché, K.B.; Frede, H.-G.; Breuer, L. Model intercomparison to explore catchment functioning: Results from a remote montane tropical rainforest. *Ecol. Model.* **2012**, *239*, 3–13. [CrossRef]

31. Lee, J.; Kim, J.; Jang, W.S.; Lim, K.J.; Engel, B.A. Assessment of Baseflow Estimates Considering Recession Characteristics in SWAT. *Water* **2018**, *10*, 371. [CrossRef]

32. Lerner, D.N.; Issar, A.S.; Simmers, I. Groundwater recharge. A guide to understanding and estimating natural recharge. In *IAH International Contributions to Hydrogeology*; Heise: Hannover, Germany, 1990; 345p.

33. Batelaan, O.; De Smedt, F. GIS-based recharge estimation by coupling surface-subsurface water balances. *J. Hydrol.* **2007**, *337*, 337–355. [CrossRef]

34. Alcalá, F.J.; Custodio, E. Spatial average aquifer recharge through atmospheric chloride mass balance and its uncertainty in continental Spain. *Hydrol. Process.* **2014**, *28*, 218–236. [CrossRef]

35. Scanlon, B.R.; Healy, R.W.; Cook, P.G. Choosing appropriate techniques for quantifying groundwater recharge. *Hydrogeol. J.* **2002**, *10*, 18–39. [CrossRef]

36. McMahon, P.B.; Plummer, L.N.; Bohlke, J.K.; Shapiro, S.D.; Hinkle, S.R. A comparison of recharge rates in aquifers of the United States based on groundwater-based data. *Hydrogeol. J.* **2011**, *19*, 779–800. [CrossRef]

37. Claasen, H.C.; Reddy, M.M.; Halm, D.R. Use of the chloride ion in determining hydrologic-basin water budgets: A 3-year case study in the San Juan Mountains, Colorado, USA. *J. Hydrol.* **1986**, *85*, 49–71. [CrossRef]

38. Dettinger, M.D. Reconnaissance estimates of natural recharge to desert basins in Nevada, USA, by using chloride-balance calculations. *J. Hydrol.* **1989**, *106*, 55–78. [CrossRef]

39. Wood, W.W.; Sanford, W.E. Chemical and isotopic methods for quantifying ground-water recharge in a regional, semiarid environment. *Ground Water* **1995**, *33*, 458–468. [CrossRef]

40. Sami, K.; Hughes, D.A. A comparison of recharge estimates to a fractured sedimentary aquifer in South Africa from a chloride mass balance and an integrated surface-subsurface model. *J. Hydrol.* **1996**, *179*, 111–136. [CrossRef]

41. Scanlon, B.R.; Keese, K.E.; Flint, A.L.; Flint, L.E.; Gaye, C.B.; Edmunds, W.M.; Simmers, I. Global synthesis of groundwater recharge in semiarid and arid regions. *Hydrol. Process.* **2006**, *20*, 3335–3370. [CrossRef]

42. Alcalá, F.J.; Custodio, E. Natural uncertainty of spatial average aquifer recharge through atmospheric chloride mass balance in continental Spain. *J. Hydrol.* **2015**, *524*, 642–661. [CrossRef]

43. Paz, C.; Alcalá, F.J.; Carvalho, J.M.; Ribeiro, L. Current uses of ground penetrating radar in groundwater-dependent ecosystems research. *Sci. Total Environ.* **2017**, *595*, 868–885. [CrossRef]

44. Capel-Molina, J.J. *Los Climas de España*; Oikos-Tau: Barcelona, Spain, 1981; 403p.

45. Trigo, R.; Pozo-Vázquez, D.; Osborn, T.; Castro-Díez, Y.; Gámiz-Fortis, S.; Esteban-Parra, M. North Atlantic oscillation influence on precipitation, river flow and water resources in the Iberian peninsula. *Int. J. Climatol.* **2004**, *24*, 925–944. [CrossRef]

46. Vanderlinden, K.; Giraldez, J.V.; Van Meirvenne, M. Assessing Reference Evapotranspiration by the Hargreaves Method in Southern Spain. *J. Irrig. Drain. Eng.* **2004**, *130*, 184–191. [CrossRef]

47. Azéma, J.; Foucault, A.; Fourcade, E.; García–Hernández, M.; González–Donoso, J.M.; Linares, D.; López–García, A.C.; Rivas, P.; Vera, J.A. *Las Microfacies del Jurásico y Cretácico de las Zonas Externas de las Cordilleras Béticas*; Servicio de Publicaciones de la Universidad de Granada: Granada, Spain, 1979.

48. Vera, J.A. *Geología de España*, 1st ed.; Sociedad Geológica de España e Instituto Geológico y Minero de España, Ministerio de Educación y Ciencia: Madrid, Spain, 2004; 884p.

49. Sanz de Galdeano, C.; Peláez, J.A. *La Cuenca de Guadix–Baza. Estructura, Tectónica Activa, Sismicidad, Geomorfología y Dataciones Existentes*; Universidad de Granada–CSIC: Granada, Spain, 2007; 351p.

50. IGME. Hydrogeological Map of Spain, Scale 1:200,000; Sheet n° 78, Baza; Geological Survey of Spain, Memory and Maps. 1988. Available online: http://info.igme.es/cartografiadigital/tematica/Hidrogeologico200.aspx (accessed on 20 January 2020).

51. IGME. *Hydrogeological Map of Spain, Scale 1:200,000*. Sheet n° 71, Villacarrillo; Geological Survey of Spain, Memory and Maps. 1995. Available online: http://info.igme.es/cartografiadigital/tematica/Hidrogeologico200. aspx (accessed on 20 January 2020).

52. IGME. *Proyecto para la actualización de la infraestructura hidrogeológica de las Unidades 05.01 Sierra de Cazorla, 05.02 Quesada-Castril, 07.07 Sierras de Segura-Cazorla y el Carbonatado de la Loma de Úbeda*; Geological Survey of Spain and General Directorate for Water Planning; Ministry of Industry: Madrid, Spain, 2001. (In Spanish)

53. Peralta-Maraver, I.; López-Rodríguez, M.J.; de Figueroa, J.T. Structure, dynamics and stability of a Mediterranean river food web. *Mar. Freshwater Res.* **2017**, *68*, 484–495. [CrossRef]

54. Alcalá, F.J.; Custodio, E. Atmospheric chloride deposition in continental Spain. *Hydrol. Process.* **2008**, *22*, 3636–3650. [CrossRef]

55. Alcala, F.J.; Custodio, E. Using the Cl/Br ratio as a tracer to identify the origin of salinity in aquifers in Spain and Portugal. *J. Hydrol.* **2008**, *359*, 189–207. [CrossRef]

56. Hurrell, J.W. Decadal trends in the North Atlantic Oscillation, regional temperatures and precipitation. *Nature* **1995**, *269*, 676–679. [CrossRef]

57. Raposo, J.R.; Dafonte, J.; Molinero, J. Assessing the impact of future climate change on groundwater recharge in Galicia-Costa, Spain. *Hydrogeol. J.* **2013**, *21*, 459–479. [CrossRef]

58. Barberá, J.A.; Jódar, J.; Custodio, E.; González-Ramón, A.; Jiménez-Gavilán, P.; Vadillo, I.; Pedrera, A.; Martos-Rosillo, S. Groundwater dynamics in a hydrologically-modified alpine watershed from an ancient managed recharge system (Sierra Nevada National Park, Southern Spain): Insights from hydrogeochemical and isotopic information. *Sci. Total Environ.* **2018**, *640–641*, 874–893. [CrossRef]

59. Pulido-Velazquez, D.; Collados-Lara, A.J.; Alcalá, F.J. Assessing impacts of future potential climate change scenarios on aquifer recharge in continental Spain. *J. Hydrol.* **2018**, *567*, 803–819. [CrossRef]

60. Nathan, R.J.; McMahon, T.A. Evaluation of automated techniques for base-flow and recession analyses. *Water Resour. Res.* **1990**, *26*, 1465–1473. [CrossRef]

61. Arnold, J.G.; Allen, P.M.; Muttiah, R.; Bernhardt, G. Automated Base Flow Separation and Recession Analysis Techniques. *Ground Water* **1995**, *33*, 1010–1018. [CrossRef]

62. Meaurio, M.; Zabaleta, A.; Angel, J.; Srinivasan, R.; Antigüedad, I. Evaluation of SWAT models performance to simulate streamflow spatial origin. The case of a small forested watershed. *J. Hydrol.* **2015**, *525*, 326–334. [CrossRef]

63. Arnold, J.G.; Allen, P.M. Automated Methods for Estimating Baseflow and Ground Water Recharge from Streamflow Records. *JAWRA J. Am. Water Resour. Assoc.* **1999**, *35*, 411–424. [CrossRef]

64. Senent-Aparicio, J.; Pérez-Sánchez, J.; Carrillo-García, J.; Soto, J. Using SWAT and Fuzzy TOPSIS to Assess the Impact of Climate Change in the Headwaters of the Segura River Basin (SE Spain). *Water* **2017**, *9*, 149. [CrossRef]

65. Jimeno-Sáez, P.; Senent-Aparicio, J.; Pérez-Sánchez, J.; Pulido-Velazquez, D. A Comparison of SWAT and ANN models for daily runoff simulation in different climatic zones of peninsular Spain. *Water* **2018**, *10*, 192. [CrossRef]

66. Abbaspour, K.C.; Rouholahnejad, E.; Vaghefi, S.; Srinivasan, R.; Yang, H.; Kløve, B. A continental-scale hydrology and water quality model for Europe: Calibration and uncertainty of a high-resolution large-scale SWAT model. *J. Hydrol.* **2015**, *524*, 733–752. [CrossRef]

67. Moriasi, D.N.; Wilson, B.N.; Douglas-Mankin, K.R.; Arnold, J.G.; Gowda, P.H. Hydrologic and water quality models: Use, calibration, and validation. *Trans. ASABE* **2012**, *55*, 1241–1247. [CrossRef]

68. Hargreaves, G.H.; Samani, Z.A. Estimating potential evapotranspiration. *J. Irrig. Drain. Div. ASCE* **1982**, *108*, 225–230.

69. Arnold, J.G.; Kiniry, J.R.; Srinivasan, R.; Williams, J.R.; Haney, E.B.; Neitsch, S.L. Soil and Water Assessment Tool—Input/Output Documentation—Version 2012. Available online: http://swat.tamu.edu/documentation/ (accessed on 20 December 2019).

70. Moral, F.; Cruz-Sanjulian, J.J.; Olias, M. Geochemical evolution of groundwater in the carbonate aquifers of Sierra de Segura (Betic Cordillera, Southern Spain). *J. Hydrol.* **2008**, *360*, 281–296. [CrossRef]

71. Senent-Aparicio, J.; Jimeno-Sáez, P.; Bueno-Crespo, A.; Pérez-Sánchez, J.; Pulido-Velazquez, D. Coupling machine-learning techniques with SWAT model for instantaneous peak flow prediction. *Biosyst. Eng.* **2019**, *177*, 67–77. [CrossRef]

72. Krause, P.; Boyle, D.P.; Base, F. Comparison of different efficiency criteria for hydrological model assessment. *Adv. Geosci.* **2005**, *5*, 89–97. [CrossRef]

Article

The Ecosystem Resilience Concept Applied to Hydrogeological Systems: A General Approach

África de la Hera-Portillo [1,*], Julio López-Gutiérrez [1], Pedro Zorrilla-Miras [2], Beatriz Mayor [2] and Elena López-Gunn [2]

[1] Geological Survey of Spain/Instituto Geológico y Minero de España (IGME), 28003 Madrid, Spain; j.lopezgu@igme.es

[2] ICATALIST, Borni 20, 28232 Madrid, Spain; pzorrilla-miras@icatalist.eu (P.Z.-M.); bmayor@icatalist.eu (B.M.); elopezgunn@icatalist.eu (E.L.-G.)

* Correspondence: a.delahera@igme.es; Tel.: +34-91-3495-967

Received: 30 May 2020; Accepted: 22 June 2020; Published: 25 June 2020

Abstract: We have witnessed the great changes that hydrogeological systems are facing in the last decades: rivers that have dried up; wetlands that have disappeared, leaving their buckets converted into farmland; and aquifers that have been intensively exploited for years, among others. Humans have caused the most part of these results that can be worsened by climate change, with delayed effects on groundwater quantity and quality. The consequences are negatively impacting ecosystems and dependent societies. The concept of resilience has not been extensively used in the hydrogeological research, and it can be a very useful concept that can improve the understanding and management of these systems. The aim of this work is to briefly discuss the role of resilience in the context of freshwater systems affected by either climate or anthropic actions as a way to increase our understanding of how anticipating negative changes (transitions) may contribute to improving the management of the system and preserving the services that it provides. First, the article presents the basic concepts applied to hydrogeological systems from the ecosystem's resilience approach. Second, the factors controlling for hydrogeological systems' responses to different impacts are commented upon. Third, a case study is analyzed and discussed. Finally, the useful implications of the concept are discussed.

Keywords: ecosystems; hydrogeological system; sustainability; significant damage; resilience

1. Introduction

Groundwater and surface water resources are heavily exploited in many parts of the world, and freshwater demands are increasing globally [1]. Any alteration of the baseline conditions of the system may lead to an undesirable state (degradation of quality or quantity). Anthropogenic effects disturb the natural processes of aquifers, and the equilibrium within the unsaturated and saturated zones, and may also increase the contents of undesired substances in groundwater. There are different types of impact affecting hydrogeological systems at different scales. The impacts or disturbances may be either natural or anthropic. Natural disturbances include any type of catastrophe that can affect a hydrogeological system, such as earthquakes, climate (extreme events but also climate variability), or fires. Anthropic disturbances include pumping and various polluting activities such as discharges; agricultural, industrial and nuclear activities; the filtration of substances stored underground; injection into wells; and urban solid waste deposits, among others. Groundwater quality and quantity degradation owing to intensive aquifer exploitation is recorded in many countries [2–8].

This article aims to contribute to develop a better understanding of the concept of resilience when applied to hydrogeological systems, which, in turn, will help develop a better understanding of the buffering capacity of hydrogeological systems. This represents a step to be able to anticipate the potential impacts on the system of specific changes and the system's response. This requires focusing attention on the internal variables which are more sensitive to impacts. The concept of resilience can be helpful to avoid these problems. For example, based on the established groundwater baseline patterns, changes can be identified from the very beginning and can provide an early warning signal to make decisions on sustainable groundwater management. Resilience is also related to water security as groundwater is regarded as one of the most reliable yet also vulnerable sources of drinking water in many countries. This is important since the substantial decline of groundwater levels may affect the water security of a growing economy [9].

There exists an extensive literary record dealing with the resilience of natural systems to different impacts, although most of this literature does not deal with resilience explicitly. Until very recently, the term "resilience" did not appear in hydrogeology glossaries. The idea of the resilience of a "hydrogeological space" or "hydrogeological medium" was developed by LV Demidyuk, NI Lebedeva and GA Golodkovskaya in the 1970s and 1980s, but unfortunately these reports were published in Russian only, e.g., Golodkovskaya and Eliseyev (1989) [10]. There are barely 20 publications in the Web of Science (WOS) returned by the search terms "resilience" and "hydrogeology", and most of these are articles do not treat resilience as a central topic [11–20]. In the field of hydrogeology, the most frequent works dealing with resilience are specific and local, and the concept of resilience is not approached from a generic point of view. Our work aims to contribute to these conceptual reflections.

There is often some confusion in the literature regarding the application of the concept of resilience. Some works apply the concept to groundwater or water resources (liquid phase), and others to aquifers (physical environment). This is an important difference as the intrinsic properties of the system vary in each case. The confusion stems from the fact that the descriptor variable most frequently selected is the same: the piezometric level.

Most literature uses aquifer resilience (AR) as a conclusion derived from the research (Cuthbert et al. (2019) [11] Maurice et al. (2019) [12], Mazi et al. (2014) [14], Chinnasamy et al. (2018) [16], De Eyto et al. (2016) [17], Hejazian et al. (2017) [21]), with the largest number of works focused on the analysis of drought as a disturbing element (Lorenzo-Lacruz et al. (2017) [13], McDonald et al. (2017) [22]. Two references present a deeper study of the resilience of aquifers. First, Bouska et al. (2019) [23] apply the concept of general resilience to the restoration of large river ecosystems in the Upper Mississippi. However, their approach is more biological [24,25] than hydrogeological, i.e., they develop indicators for three principles of general resilience: diversity and redundancy, connectivity, and controlling variables. The latter includes historical water level fluctuations, water clarity, nutrient concentration, and invasive aquatic species. Wurl et al. (2018) [26] adopt an approach that is closer to the resilience concept from a hydrogeological perspective. The authors designed and used a set of indicators as outcomes for combined human–water systems to predict water trajectories under different human impacts.

This literature review identifies a number of tools that have been used indirectly to build the application of resilience into hydrogeology: the use of tracers or isotopes to determine groundwater age [27], the analysis of piezometric evolution trends in aquifers [28], and the quantification of water consumption for agriculture in restricted aquifers [29]. Quantification also relies on qualitative indicators as an auxiliary tool [16]. However, the term "resilience" in the hydrogeological literature is presented as an attribute that is not analyzed in itself, but which is frequently cited within the framework of processes (recharge, precipitation) associated mostly with climatic variability, and, to a much lesser degree, with other processes (extreme events).

The aim of this work is to increase our understanding of how hydrogeological systems deal with disturbances as a way to anticipate transitions (changes in the system affecting their functioning) and to propose a conceptual model for the analysis of the resilience of aquifers. In turn, this knowledge can support the more sustainable management of the system. The paper compiles the key knowledge around this concept that can be applied to hydrogeological systems and discusses a relevant case for illustrative purposes.

The article is structured as follows: first, the article presents the basic concepts applied to hydrogeological systems, and the conceptual model proposed, including factors controlling the responses of aquifers to different types of impacts. Second, a series of considerations are made regarding its scope. It concludes with a discussion and considerations of how to deal with the resilience property in the framework of long-term data series to obtain analytical and useful results.

2. Hydrogeological Systems and the Resilience Concept

2.1. Hydrogeological Systems as Complex Systems

Hydrogeological water systems operate with a certain behavior for a certain period at the human time scale [30–35]. For example, depending on their water regime, rivers can be classified as permanent, seasonal, temporary or intermittent; wetlands can be permanent, temporary, seasonal or erratic based on their hydroperiod; and aquifers can be classified as unconfined, confined, semi-confined or perched according to their operation.

In many areas of the planet, hydrogeological systems (water systems) are subject to different types of stressors, such as the extraction of water for irrigated agriculture, or in urban coastal areas contamination from localized or extended sources and changes in recharge regimes due to climate change, which may cause the systems to exceed the limits of sustainability [36]. If this occurs, the systems become unbalanced, leading to a tipping point [37] in their behavior, bringing them to a new state of equilibrium [38].

2.2. The Resilience Concept and Theory

Resilience is fundamentally directed to the way a system responds to a disturbance [39,40] that can be punctual or a long-term process, for example one implying gradual alterations (slow onset changes) [41]. This concept originated in metallurgy and has subsequently been applied in many other disciplines. In the field of ecology, the concept is based on the theory that systems are in a natural state of flux rather than an equilibrium [42]. In this work, we define resilience as a system's ability to recover a situation of equilibrium or metastability (known as state, see the definition below), characterized by a known behavior. We do not see it as a return to its pristine conditions for two reasons: (a) in many areas, there is no information available to define the pristine conditions, i.e., it refers to a period prior to registration for which there is no data; (b) in general terms, the systems are altered (by humans or by other natural processes) in one way or another, so that their return to an pristine state may be a goal that is significantly outside the realms of possibility.

The literature on the resilience of complex systems is highly fragmented [43,44]. One of the main problems that emerges is the lack of terminological consensus among the authors [44,45]. Different authors use a range of terms to refer generically to the same concept of "complex systems", including "complex adaptive ecosystems" [46,47], "complex adaptive systems", "complex, coupled Socio-Ecological Systems (SESs)", and "complex, multi-scalar SESs" [48]. In some cases, they include references to impacts, events, shocks, pulses, threats or stress, leading to a terminological confusion that must be avoided in a scoping study such as the one presented here.

From a biological perspective, the concept of resilience is especially applicable to natural systems that adapt to different degrees of disturbance while maintaining the same processes and structures that reinforce each other [49,50], and whose connection is known as a "regime" [51] (Table 1). The transition from one regime to another (regime shift) occurs through thresholds (Figure 1); and the new regime is characterized by a different set of processes and structures (behavior) [52]. Regime changes are typically associated with significant consequences in processes or structures (e.g., a change in water composition that leads to a loss of water quality), and do not always occur in sudden leaps or at turning points in their trajectory, but may be the result of long system periods [53] or slow and progressive changes [41]. Not all regime changes entail a tipping point. Indeed, Bertalanffy (1968) [54] identifies six mechanisms that can trigger a regime change (slow–fast cyclic transition, stochastic resonance, noise-induced transition, long transient upon extreme events, big stepwise changes in drivers), but only one of them involves a tipping point: slowly changing driver to tipping point. Four out of these six transition indicators can be used as early-warning signals [37,55,56]. In the field of hydrogeological ecosystem research, the aspects of the dynamics of changes between the states of equilibrium are relatively unexplored.

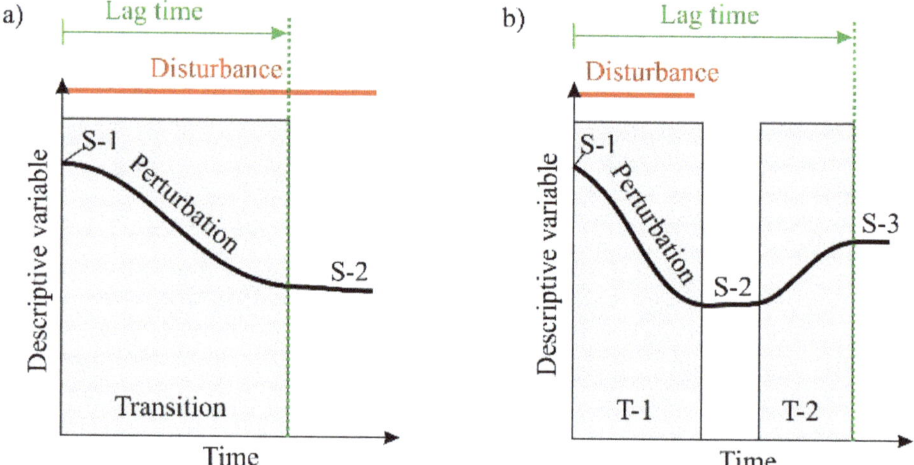

Figure 1. Scheme of the aquifer resilience (AR) through the temporal evolution of a descriptive variable. The black lines indicate the time series observed of the aquifer system, and the red lines represent the time series of the underlying environmental conditions. (**a**) In a theoretical example, the pumping of groundwater from an aquifer (disturbance) prolonged over time will cause a water table depletion (parameter) which could reflect the aquifer response to the impact. (**b**). If the pumping ceases (disturbance stops), the aquifer may recover a new equilibrium (State-3). Both behaviors (**a**,**b**) are useful to understand aquifer resilience and must be taken into account in the interpretation of the AR (Source: authors' own).

Table 1. Extended definition of terms associated with resilience in a biological environment. Text in italics indicates the literal definition of the authors. Text in brackets are comments from the authors. When the source is not indicated, the reference is proper.

Word	Definition	Source
Adaptive capacity	Latent potential of an ecosystem to alter resilience in response to change. Similarly, in the ecological sciences, adaptation, adaptedness, adaptability and adaptive capacity, terms with different meanings, have often been used interchangeably.	[57]
Alternative state/regime	A potential alternative configuration of a system in terms of the structural and functional composition, processes, and feedbacks.	[57]
Critical slowing down (CSD)	CSD occurs as the system approaches the threshold, the distance to the critical threshold is reduced, the recovery rate decreases and ecological resilience declines.	[37]
Early-warning signal (EWS)	A statistical signal indicative of a system approaching a critical transition. (Often used interchangeably with leading indicator. Examples are variance or autocorrelation.)	[56] (p. 906)
Forcing	External pressures that destabilize a system, pushing it towards a tipping point.	[56] (p. 906)
Hydrogeological systems	Set of geological formations whose hydrogeological functioning should be considered together.	
Linear system	System whose behavior is expressible by adding the behaviors of its descriptors.	

Table 1. *Cont.*

Word	Definition	Source
Perturbance/ disturbances	Alteration in the order or the permanent characteristics that comprise the normal development of a process.	
Pressure	Activities subject to generate impacts on groundwater	[58]
Regime shift ("change" in other references)	Persistent change in structure, function, and feedback of an ecosystem. (This term is used interchangeably with "critical transition" in the literature.) We will use "state" instead of "regime" in this paper.	[57]
Scale	The geographic extension over which a process operates and the frequency with which a process occurs.	[1,51]
Stability	A system characteristic whereby system properties remain unchanged following disturbance. Adaptive capacity can increase stability, but system components can fluctuate (and are therefore unstable) while still remaining within the range of values that signify a particular state.	[57]
Stressor	Stimuli or situations capable of producing certain changes that trigger the stress response.	
Transient regime	Response of a system that changes over time, as opposed to the permanent regime.	
Tipping point (threshold, bifurcation point)	The point at which a system is so unstable that even small perturbations cause dramatic shifts in its state.	[59]
Variables describing change (fast variables, controlling variables and control variables)	Fast and slow variables. "Fast" variables are those that are of primary concern to system users. The dynamic of these fast variables is strongly shaped by other system variables that generally change much more slowly. "Slow" variables or controlling variables are not the same as control variables.	[45]

The scope of the concept of ecosystems resilience is broader than initially considered. When discussing ecosystems alone, resilience is closely related to sustainability ([60,61]). Scheffer et al. (2001) [62] report that a loss of ecosystem resilience generally paves the way for a change to an alternative state and suggest that sustainable management should be directed towards maintaining ecosystem resilience [62].

Resilience also refers to the system's adaptive capacity [57] and vulnerability, given that it offers another approach to the changes produced by a disturbance; however, this idea is controversial, as some authors consider that the concept of adaptive capacity is muddled with multiple meanings in current use often being indistinguishable from resilience [41]. Resilience is an intrinsic property of the system that emerges from certain changes. Some authors define the adaptive capacity of ecosystems as a latent potential quality to alter resilience in response to change [57].

Resilience has implications in the socio-economic and political spheres, since knowing the dimensions of this attribute enables managers to intervene in the natural environment before a change of regime or favoring one state of equilibrium over another.

Although the term "resilience" is increasingly used by political and environmental managers, it remains vague, variable and difficult to quantify [39]. This work clarifies what it means from the hydrogeological perspective, without attempting to review the state of the art of the concept or to list an inventory of works in which the concept is applied to geological and hydrogeological studies.

2.3. Resilience from Ecology to Hydrogeology: A Conceptual Framework for Its Analysis

The analysis of resilience focuses on the dynamics of the system, particularly looking at two areas: the cause effect (disturbance), and its consequence in the system (system response to the disturbance). The disturbance leads the system to alter certain internal variables, which define the new equilibrium state (regime). Based on the concept of resilience and on the literature dealing with the resilience of hydrogeological systems, we propose a conceptual model that can use aquifer resilience to support its management (Figure 2).

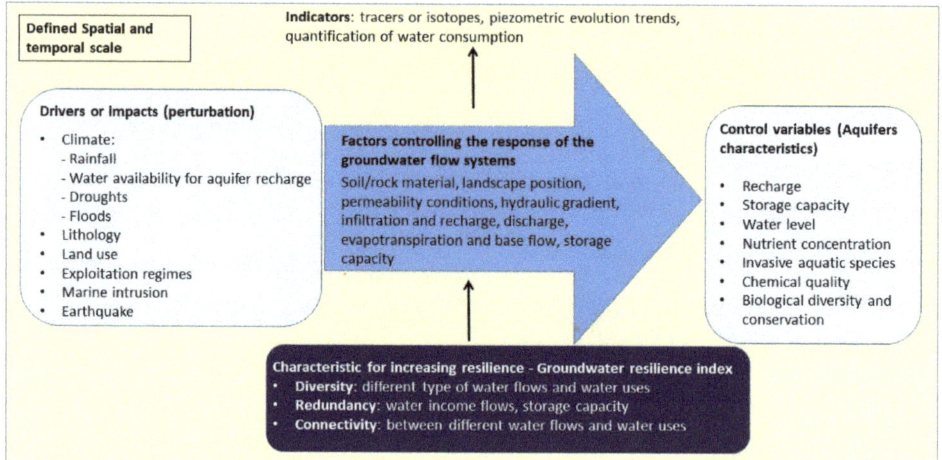

Figure 2. Proposed conceptual model for using the resilience concept in the management of aquifers (Source: authors' own).

Step 1. The first step must be the system definition, including its geological boundaries and its main characteristics, flows and functions. An important issue to be addressed in any resilience analysis is the scale of work (space–time dimensions).

To understand hydrogeological system resilience, cause–effect relationships, and impacts, it is necessary to have information about its functioning. Groundwater flow systems depend on both the hydrogeological characteristics of the soil/rock material and on the landscape position [3]. These factors control the permeability conditions and the hydraulic gradient differences which regulate the groundwater flow movement. Not every impact affecting the Groundwater Flow System (GWFS) affects its hierarchical structure and functioning; this will depend on the nature, magnitude and duration of the impact, and the factors controlling the response of GWFS to those pressures.

Step 2. The second step is to describe the process triggered by the disturbance (some authors refer to them as descriptive variables of the change and the interactions between them) that is, to describe its magnitude, duration and scale. This involves monitoring the variables, describing the change before, during and after the change. The second step is therefore to define and describe the disturbance that acts on the system (state 1) causing a series of internal changes that lead to a situation of instability for a certain or indefinite time (Figure 1). The usual process is for systems to tend to a new state of equilibrium (state 2) through internal changes and interactions between the variables that describe the change.

In describing the disturbance, there are a set of elements that need to be considered when analyzing resilience. One of them is the time scale of the system and of the disturbing forces. In the natural environment, some internal changes occur over short time periods and are visible on a human time scale (for example, change in the eutrophication conditions of a lake, reduction in the population of a certain insect, etc.). In other cases, the effect of a disturbing agent may not become evident for years, millennia or millions of years, and thus be difficult to determine, for example given the different temporal scales of geological processes.

Another element to consider is the possible overlap of effects due to the vast dimensions of a system and the difference in the periodicity and breadth of the various antagonistic processes. For example, in a large detrital aquifer with an immense storage capacity, a short extremely dry period can be obliterated by the hyper annual natural recharge of average and humid years, i.e., the overlapping of the previous and subsequent average recharge would have cushioned these effects.

Step 3. The third step is to describe the new state of equilibrium (regime) [45].

The analysis of resilience focuses on the disturbance, the processes of change, and the system's recovery into a new state of equilibrium. If the concept of resilience is to be made operational, we need to find ways to measure it. Therefore, an element to consider is the fact that the transition from one state (or "alternative state" according to some authors) to another occurs through thresholds [63]. Some authors argue that these thresholds are crucial for measuring resilience as these offer a way to quantify how much disturbance can be absorbed by a system before switching to another regime. The identification of these thresholds requires experimental or observational data on the changes between regimes in a certain system and, if possible, on its recovery trajectory [39]. Although there is considerable work done on transition indicators in biological systems [56], this is not the case for geological systems. A long time series of data is not always possible. In these cases, there are other resources to carry out this analytical study, as we show in the next section.

3. Conceptual Model Applied to a Real Case: The Upper Guadiana Basin in Central-West Spain

The case study presented here is for illustrative purposes to show through an example why the lens of resilience is valuable to gain a better, more anticipative knowledge of the system. It is also a pertinent case because it is applied to the functioning of a very large aquifer, its relation with an important groundwater-dependent wetland (the Tablas de Daimiel National Park, TDNP), and the dynamic of the system from a starting point, to highly degraded systems (both aquifer and wetland), and to a current significant recovery.

In the following sections, a description of the main steps mentioned above is presented. Figure 3 summarizes the main factors considered.

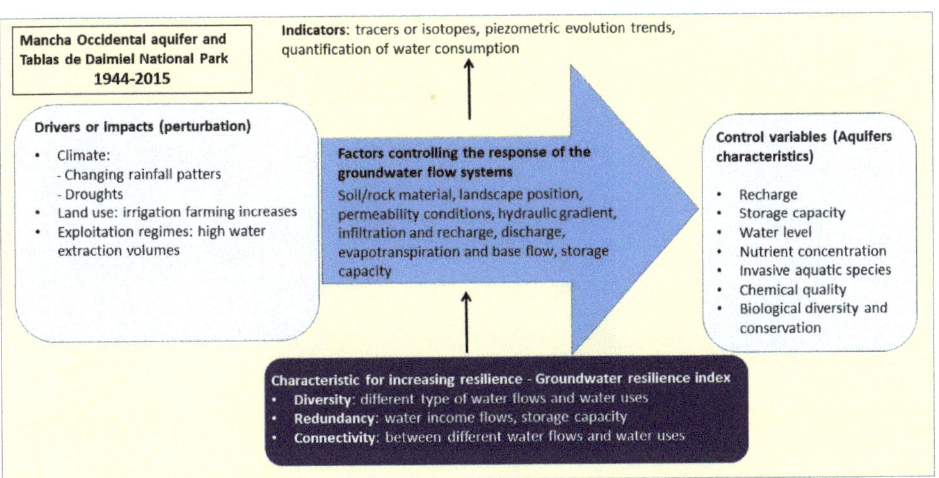

Figure 3. Conceptual model applied to the Upper Guadiana Basin (Source: authors' own).

3.1. First Step: Description of the System, Flows and Functions

This case provides, on the one hand, the aquifer perspective through the Mancha Occidental aquifer, and, on the other hand, the wetland perspective represented by a groundwater-dependent ecosystem affected by multiple impacts with enough data to allow its analysis. The equilibrium in wetland ecosystems is very fragile, showing high sensibility and vulnerability [64–67].

Tablas de Daimiel is a groundwater-dependent ecosystem subject to different types of impacts, both climatic and anthropogenic. This wetland is the main discharge outlet of the Upper Guadiana basin's aquifers (Figure 4), in such a way that it can be considered the "thermometer" of the 16,000 km^2 groundwater system [4]. The Tablas de Daimiel wetland has existed for over 250,000 years, evolving from a deep lake to a fluctuating shallow system, with different reversible intermediate phases depending on hydroclimatic conditions [8]. In 1960, the system water inflows combined brackish surface water from the Cigüela River with freshwater inputs from the Guadiana River and the underlying aquifer.

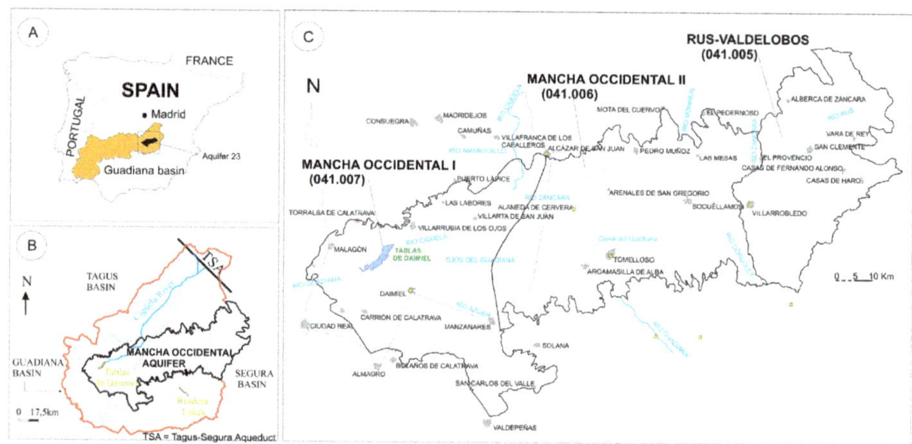

Figure 4. (**A**) General geographical setting. (**B**) Upper Guadiana basin. (**C**) The Mancha Occidental Aquifer (currently divided into three groundwater bodies: Mancha Occidental I, Mancha Occidental II and Rus-Valdelobos) (Source: Authors' own).

The analysis of the flood data series for the period 1944–1974 (Figure 5), prior to the overexploitation of the aquifer, reveals that in that period changes in rainfall (Figure 6) determined changes in water variability: changes in rainfall determined changes in the wetland surface covered by water [68].

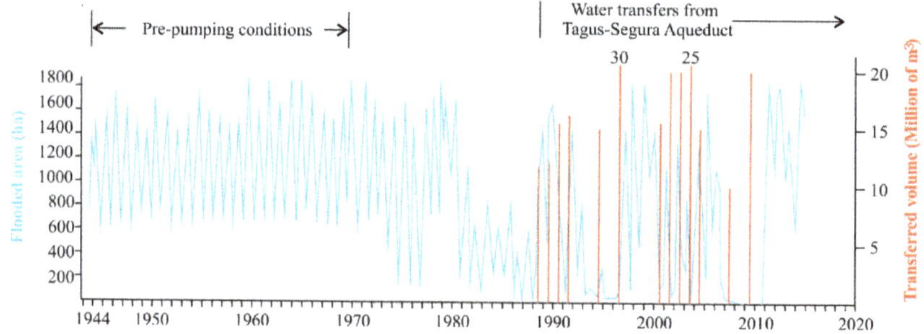

Figure 5. Flooded area evolution of Tablas de Daimiel National Park. Data from Sánchez-Carrillo et al. 2016 [68]. At the end of May 2020, the flooded area is less than 80 ha, and a new water transfer has been requested by the Tablas de Daimiel National Park (TDNP) Director to the Government (Source: Authors' own).

In resilience terms, 1944–1970 corresponds to the pre-pumping stage of the system and can be used as a reference stage for our purpose.

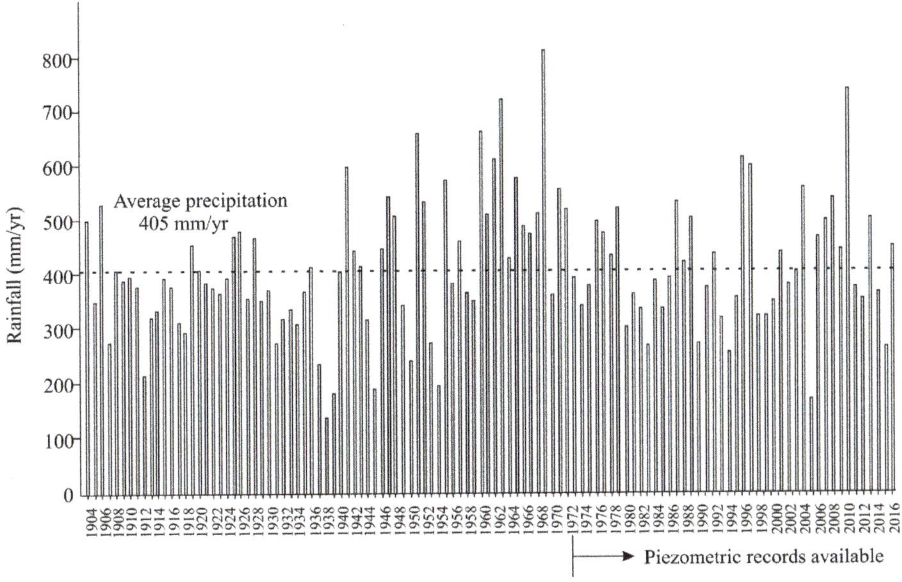

Figure 6. Long-term rainfall patterns in Mancha Occidental Aquifer 1904–2014. Rainfall data from weather stations 4121 and 4121C (data from [8]) (Authors' own).

3.2. Water Quality in Tablas de Daimiel National Park: Baseline Stage (September–October 1974)

Electrical conductivity ranged between 800 and 1600 μS/cm, with a minimum of 300 and a maximum of 5400 μS/cm. The dominant anions were sulphate and bicarbonate, whereas the dominant cations were calcium and magnesium [69]. Calcium bicarbonate waters predominated to the northwest and northeast of the national park, as well as in the vicinity of the Ojos del Guadiana springs, with the lowest conductivity sampled. Meanwhile, calcium sulphate waters predominated around the left bank of the wetland [4].

3.3. Second Step: Description of the Perturbation

Since the mid-1970s, the intensive exploitation of the aquifer for agricultural irrigation caused the desiccation of the wetland and neighboring springs. The cause of the hydrological situation of the Tablas de Daimiel in the 1980s can be well explained due to the length of the time series covering period 1975–2008. Extensive descriptive publications exist about this period examining its origin, reasons and social-ecological consequences [5–7].

Based on data provided by Aguilera and Moreno (2018) [69], the impact of drought caused a decrease in groundwater levels, which, at the same time, produced the burning of peat underlying the wetland, causing a smoldering peat fire in 2009 [69]. In this case, the descriptive variables observed were soil moisture, temperature and organic matter content for the period 2006–2010 [70]. Continuous soil moisture and temperature monitoring is recommended as an indicator of potential combustion and auto-ignition fire risk but does not work as alert system for an already active fire. In fact, the presence of active smoke columns is a late warning. A new fire means to arrive late. Fire modifies irreversibly the physical structure of affected soils, which implies a damage to the ecosystem.

In resilience terms, this analysis shows how the system faces the burning of peat impact. This provides the added value of the reaction to the change in the hydrological conditions of the soil.

3.4. Third Step: Description of the New State of Equilibrium

The longest water table records show the evolution of the system through the different described impacts (Figure 7). With depleted piezometric levels, the TDNP wetland operates as a recharge system for a local shallow perched multi-layer aquifer disconnected from the deeper regional groundwater flow towards main irrigation areas [71]. Water-table records show the tendency of the system to behave in a roughly similar manner across its entire extension.

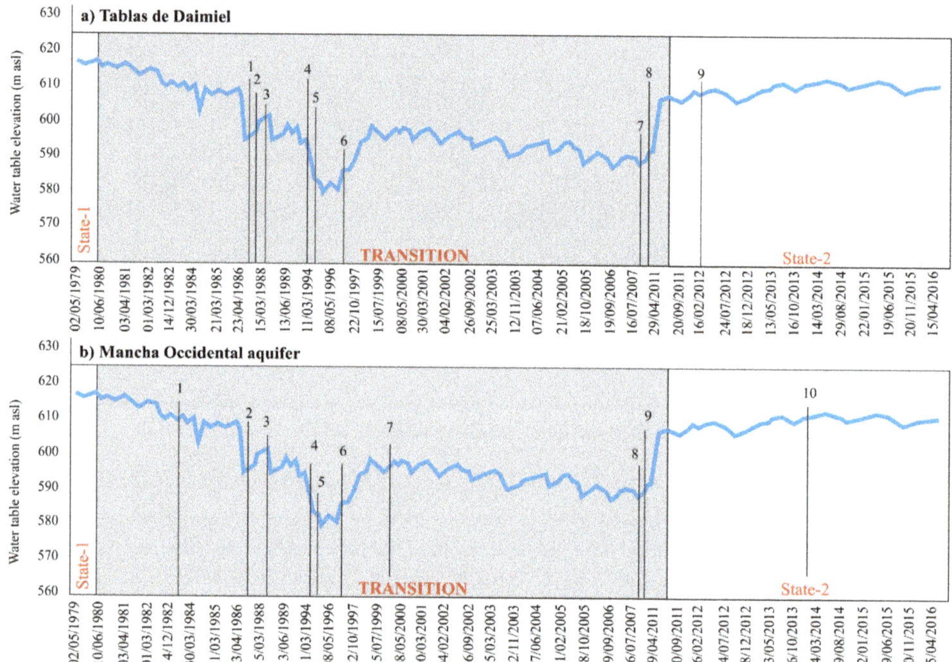

Figure 7. Water table evolution in a representative monitoring well (1930.4.0040) located near the Ojos del Guadiana springs and Tablas de Daimiel National Park. It ranks among the longest available piezometric records. The figure also shows the main historical events occurred in the (**a**) Tablas de Daimiel National Park: 01/1986–87: peat fires; 02/1987: Tablas de Daimiel was dry up for the first time; 03/1988: the first water transfer from Tagus-Segura Aqueduct; 04/1994: peat fires; 50/1995: flooded area, 30 out of 2000 ha; 06/1996–98: heavy rains; 07/2009: peat fires; 08/2010: water transfer from Tagus-Segura Aqueduct; 09/2012: groundwater feed Tablas de Daimiel. (**b**) Mancha Occidental Aquifer: 01/1983: Ojos del Guadiana springs dried; 02/1987: provisional declaration of aquifer overexploitation; 03/1988: maximum pumping peak (570 million m^3); 04/1994: declaration of aquifer overexploitation (pumping regulation); 05/1995: water level depth 47 m in Ojos del Guadiana springs; 06/1996–98: heavy rains; 07/2000: Water Framework Directive (2000/60/EC); 08/2009: water level depth 35 m in Ojos del Guadiana springs; 09/2010: heavy rains; 10/2014: Declaration of the Upper Guadiana basin's groundwater bodies (GWBs) at risk of not achieving the good quantitative and chemical status. (Data from [7]) (Authors' own).

For three decades, the wetland remained in precarious hydrological conditions, with the only exception of rapid floods due to extreme rainfall events and sporadic water transfers from the Tagus river basin. The water transfers from the Tajo-Segura Aqueduct started in 1988 and have often been carried out during spring or summer when evaporation and infiltration rates are highest and when increased demands of water for irrigation promote illegal extractions. Flooding the wetland with pumped groundwater was a management tool used constantly during dry periods to keep a minimum flooded area. Both management tools induce quantitative and qualitative impacts on the system [69].

There are several stressors acting simultaneously on the Tablas de Daimiel National Park: droughts (drying) and flooding situations, pumping, fires, treated wastewater, water transfer from other basins, and land use changes, among others. It is not possible to isolate just one cause and one effect for the analysis.

3.5. Water Quality in Tablas de Daimiel National Park: Pumping Stage

In 2007, the number of analyses is smaller than in the other two stages due to the water table dropping below the bottom of some piezometers and all springs drying up [7]. Electrical conductivity ranged between 2000 and 11,000 µS/cm. In the vicinity of the wetland, the main cations evolved into sodium and magnesium, while the dominant anions leaned towards sulphate and chloride. The surface water exerts little influence on groundwater chemistry across most of the system.

In the case study, an unusual situation has recently occurred: in 2011, a decrease in groundwater abstraction and an extraordinary wet period reversed the trend. Following a wet period (2006–2009) capped by an exceptionally humid year (2010), the aquifer experienced an unexpected recovery of groundwater levels (almost 20 m in some areas), restoring groundwater discharge to springs and wetlands, which came back to life for the first time since the early 1980s (Figure 7). For the sake of brevity, this history may be found in [8,68]; for a more recent perspective and hydrogeological functioning, details are presented in Castaño et al. (2018) [7] and Martínez-Santos et al. (2018) [8]. Here, we just mention those aspects relevant to the aim of the paper.

3.6. Water Quality in Tablas de Daimiel National Park: Restoring Stage

Data from 2014 show that the hydrological recovery has not yet been mirrored by a similar recovery in water quality. In fact, the groundwater is more saline than it used to be before the 1970s, and the predominant hydrochemical facies have shifted with meaningful spatial gradients [7]. The descriptive variable is the water quality of the groundwater around the wetland considering three stages: (i) prior to degradation of the wetland (1974, baseline); (ii) during a period of major degradation (2007); and (iii) after the most important recovery on record (2014). The pictures of Stiff diagrams obtained for each one of these milestones respond to a different state of the wetland, in this case, without data of transition among them, which is key to identify the beginning of the changings and provides the most valuable information to make responsible management decisions. In spite of this, the second picture (2007) shows an important change in water quality which indicates a remarkable internal change in the ecosystem after 33 years of pumping. The 2014 hydrochemical data indicate that the hydrological recovery of the system refers exclusively to the water balance and not to the water quality. This means that the lag time for water quality to reach a new equilibrium is shorter than the lag time of water levels to reach the position close to that of the 1970s. In fact, the reaction of groundwater levels to rainfall and decrease of pumping is fast (a short period of a few months), whereas the evolution of groundwater quality is much slower. Moreover, there are no studies to predict whether or not it will be reached or when. Considering the high level of hydrogeological knowledge existing in this area, including hydrogeological flow models, some research could be done along these lines in order not only to predict whether a new equilibrium will be reached, but also whether the system is able to keep such a state for a long time. Perhaps new actions will be needed in the context of sustainable management decisions.

3.7. Surface Flaming Fires and Smoldering Peat Fires

Both surface flaming fires and smoldering peat fires have been relatively frequent in the TDNP surroundings (1977, 1987 and 1991) [7,72]. In fact, most natural peatlands outside the park limits have disappeared. Smoldering peat fires have even been reported inside the park in 1986, 1987 and 1994 [67] (see Figure 6). But they occurred under relatively wet soil conditions, with a shallow water table located less than 1 m below the surface, and only affected small areas. In the 2009 fires, on the contrary, the soil moisture was much lower, and the water table was located deep below the surface, so the fires represented a much bigger problem [70]. This means that this fire posed an enormous risk for both the physical structure supporting the ecosystem and the quality of groundwater beneath it. The analysis of key parameters monitored in several locations of the TDNP at different depths shows that there was enough previous evidence to foresee the peat self-combustion and the risk that any surface fire could be transmitted to the subsoil. Data were taken in the vadose zone of the TDNP up to a depth of 2 meters at 12 points.

In resilience terms, this analysis shows that the soil's organic carbon content and moisture are two key variables in smoldering fires. The first is related to the amount of fuel available in the soil. The second represents a threshold condition, as below a certain moisture content peat can burn. This means that the peat combustion could be predictable allowing for pro-active management.

In resilience terms, this analysis shows that the system is more resilient to water quantity than to groundwater quality. This provides an added value for the assessment of conservation strategies.

4. Discussion—The Need for Good Quality Long-Term Data

From the analysis undertaken above, on understanding a complex hydrogeological system like the Upper Guadiana aquifer, through the lens of resilience we can gather new insights on the functioning of the system. For example, this wetland has been exposed to impacts that are not always evident and reversible [73]. The first signals of a change in water quality could have been detected if an adequate system of monitoring and data interpretation had been performed since 1970. An earlier intervention in the system could have avoided the degradation of water quality suffered currently in the wetland. Most of the processes triggered by global changes were not detectable in the short term; instead, it is necessary to adopt a longer decadal scale to understand their dynamics and evaluate their consequences [68].

The most frequent controlled variables in flow river systems are discharge data. In large rivers, these records are normally well registered. Nevertheless, in many areas, gauging stations do not have continuous records, making it difficult to undertake long-term series analysis. It is important to note that over recent decades, hydrological regimes have been changing at a very fast pace. Some progress has been made in extracting long-term signals of change from hydroclimatic data. However, further studies investigating long-term changes in river runoff, and focusing on the detection of underlying mechanisms and the disentanglement of their effects are needed [73].

The most frequent variable observed in hydrogeological studies are groundwater level fluctuations and periodical groundwater samples analysis. These are the variables controlled in most groundwater monitoring networks. It is important, therefore, to have good quality data records on groundwater abstractions to investigate the links between groundwater abstractions and their potentiometric surfaces to better understand future aquifer responses to climatic and anthropogenic stresses [74].

It is not easy to find continuous flow records from springs, since only a reduced number of springs have this information available. However, in order to identify flow patterns, other variables are needed, such as electrical conductivity and temperature, that would need to be monitored simultaneously and synchronically to the flow record. In wetlands, as in any other surface water body, monitoring the height of water in the wetland flooding control [68], the groundwater level in some close wells, and water samples from the wetland's water and groundwater [75] should also be recorded continuously [67].

While many groundwater and surface water flow systems have long-term operation histories, they do not have long-term series of data to assess such operation in depth. Very frequently, this is due to a lack of budget or a lack of staff and time to interpret these records. The hidden information behind those long-term series data should be extracted through long-term trends analyses from which some processes can be identified. Galassi et al. (2014) [15] studied the results of the effects of a 6.3 Mw earthquake on 6 April 2009 on the Gran Sasso karst aquifer in L'Aquila (Italy) by comparing biotic and abiotic data from two years prior to the event (1997 and 2005) and another post-event (year 2012), although not with contiguous hydrological years. This highlights the lack of data available to conduct this type of analysis.

A sufficient length of the time series is vital to be able to distinguish between different impacts, for example, natural climate variability and signals of climate change. When adopting strict criteria regarding data length and data quality, the available information probably decreases.

Different time horizons of observations and measurements can lead to different conclusions, and therefore time extrapolations are always risky. Hierarchy theory states that there is a control mechanism in the temporal order of systems and suggests that long-term processes (that operate mainly on wide spatial scales) restrict fast processes (that work on small spatial scales), which limits their degrees of freedom [76,77]. Although these concepts have theoretical strength, their empirical evidence has not been widely demonstrated so far due to the lack of data sets [68].

It is thus argued that better management decisions could be made if they were based on managing the resilience of systems rather than maintaining them as if these systems were inherently static and thus aim to return them to the statistics of, e.g., 50 years ago. In reality, natural systems are dynamic, and even more so when combined with anthropic influences, with the impacts of multiple factors of global change not present or, at least, not having the same intensity [78].

It is essential to identify the descriptive variables of the changes in order to monitor the analysis. Without adequate monitoring of these variables, it is impossible to understand the change dynamics and their scope, duration and characterization. This monitoring should be ex ante, taking a good design of the spatial observation network into account as well as an adequate periodicity of reading or sampling. Moreover, this monitoring should be permanent in order to allow data from the pre-disturbance phase to be available during and after the disturbance. Precisely, one of the reasons for the scarcity of these studies in hydrogeology is undoubtedly the absence of monitored information on geological processes. Long-term data records are required, combined with an observation network with good spatial coverage [79,80], to facilitate the analysis of the system's resilience. Furthermore, data limitations and the lack of information on mechanisms and processes pose significant limitations to research in many systems [19,75]. Some anthropogenic interventions may imply permanent or long-term durable changes, like the construction of buildings or the start of a new groundwater competing sector, such as agriculture. Permanent or long-term durable changes (in a human scale) imply that it is often difficult to return a system to its initial conditions in the temporal and socio-economic spheres. It makes more sense to talk about resilience today in terms of considering that the system attains a new state of equilibrium under the new conditions of change caused by a disturbance [79,80]. After the cessation of a disturbance, it can be possible to return to the initial state if no new disturbances occur, although this depends on the recovery capacity of the system. Nevertheless, it is important to point out that initial does not mean pristine. In cases where an initial state is known, the management could try to return the system to this initial state as opposed to the pristine state.

To date, transition indicators can be defined in certain natural systems, including volcanic (terrain changes prior to eruption) and hydrological systems (surface water and groundwater quality changes). However, many system state changes, let alone state change thresholds, can only be roughly recognized [79,80]. Also, early-warning signals are an open field of work and will be the next step once the system transitions can be identified.

The study of the resilience of natural systems requires multidisciplinary research including teams of experts in different fields. For the case of wetlands, it is necessary to know not only their characteristic functions but also the interrelation between hydrogeological and biological processes, and particularly the dynamic of governing and socio-ecological variables. The compartmentalization of the natural environment into separate disciplinary fields merely reduces the visual scope and skews the dynamics of the natural processes that develop between both spheres conceptually. Collaboration between the different disciplines enriches this vision, guarantees a more effective approach to reality, and is the safeguard of true scientific advancement.

5. Conclusions

The analysis of the resilience of any natural system must be defined based on its twofold nature, that is, from what to what, and how. The work must focus on analyzing the type of disturbance and identifying the changes produced by this disturbance in the internal dynamics of the system (the operation of the system), both within and between states.

It is important to consider the different scales of analysis and precisely define the trigger for the changes (impact) as the external agent acting from a higher scale, and the space–time scope of what is considered the "system" under study. Within this system, the variables that describe the change must be identified, along with their evolution, the interaction between them, and their potential recovery once the disturbance ceases or the shock is cushioned.

Our exemplification through the case of the case of Tablas de Daimiel describes how a set of changes have been caused by a series of impacts. These impacts act simultaneously making difficult any correlation of cause–effect binomial. Since the system is changing in a complex way, as a consequence of global changes, the only option is to obtain good quality data with a long enough time series to be able to discriminate a system's responses to different causes. The changes occurred in the Tablas de Daimiel are a response to multiple disturbances, and their interpretation varies with the time scale considered [68]. The response of many processes triggered by different changes is reflected in a time lag that is impossible to detect with short observation periods. The Tablas de Daimiel is currently a system in a new state. In this paper, some examples have been shown covering how and at what speed ecosystems have moved their structure to this new state, using flooding, rainfall and groundwater records jointly with additional short periods of specific data (water quality, and soil moisture, temperature and organic matter content). The Tablas de Daimiel shows a high resilience to droughts and flood events and a low resilience to pumping and fires. This means that the system copes better with natural disturbances than anthropic disturbances. However, no measurements or estimations of the resilience change starting point could be made due to the lack of continuous flow recording data. Thus, good data series are key to having a strong conceptual understanding of the resilience of hydrogeological systems that in turn allow for a more adaptive style of management that better reflects that systems are not static but rather are constantly evolving. It is thus critical to understand this so that system resilience is in line with the protection of key hydrogeological system functions.

Author Contributions: Conceived, designed and original draft preparation: Á.d.l.H.-P. Formal analysis and review: J.L.-G. and P.Z.-M. English style and editing: E.L.-G., B.M., P.Z.-M. All authors have read and agreed to the published version of the manuscript.

Funding: This research was funded by European Commission H2020 program: NAIAD (Nature Insurance value: Assessment and Demonstration, grant number 730497).

Acknowledgments: This work has been funded by the H2020 Program of the European Commission Project "Nature Insurance value: Assessment and Demonstration" (NAIAD) No. 730497. The views expressed are those of the authors alone. The authors wish to thank to the anonymous reviewers for their useful and constructive comments.

Conflicts of Interest: The authors declare no conflict of interest.

References

1. Hinsby, K.; Troldborg, L.; Purtschert, R.; Corcho Alvarado, J. Integrated dynamic modelling of tracer transport and long term groundwater/surface water interaction using four 30 year 3 h time series and multiple tracers for groundwater dating. In *Isotopic Assessment of Long Term Groundwater Exploitation*; Edition: IAEA-TECDOC-1507; International Atomic Energy Agency, IAEA: Vienna, Austria, 2006; pp. 73–95.

2. Llamas, M.R.; Custodio, E. *Intensive Use of Groundwater. Challenges and Opportunities*; Llamas, M.R., Custodio, E., Eds.; Fundación Marcelino Botín. A.A. Balkema: Amsterdam, The Netherlands, 2003; pp. 1–478.

3. Sophocleous, M. Environmental implications of intensive groundwater use with special regard to streams and wetlands. In *Intensive Use of Groundwater. Challenges and Opportunities*; Llamas, R., Custodio, E., Eds.; IGME, Generalitat Valenciana, Fundación Marcelino Botín. A.A. Balkema Publishers: Lisse, The Netherlands; Abingdon, UK; Exton, PA, USA; Tokyo, Japan, 2003; pp. 93–112.

4. Llamas, M.R.; Varela-Ortega, C.; De La Hera, A.; Aldaya, M.; Villarroya, F.; Martínez-Santos, P.; Blanco-Gutiérrez, I.; Carmona-García, G.; Esteve-Bengoechea, P.; De Stefano, L.; et al. The Guadiana Basin. In *The Adaptive Water Resource Management Handbook*; Mysiak, J., Henrikson, H.J., Sullivan, C., Bromley, J., Pahl-Wostl, C., Eds.; Earthscan: London, UK, 2010; pp. 103–114.

5. Zorrilla, P.; Carmona, G.; De la Hera, A.; Varela-Ortega, C.; Martínez-Santos, P.; Bromley, J.; Henriksen, H.J. Evaluation of Bayesian Networks in Participatory Water Resources Management, Upper Guadiana Basin, Spain. *Ecol. Soc.* **2010**, *15*, 12. Available online: http://www.ecologyandsociety.org/vol15/iss3/art12/ (accessed on 10 March 2020). [CrossRef]

6. Martínez Cortina, L. Marco hidrológico de la cuenca alta del Guadiana. In *Conflictos Entre el Desarrollo de las Aguas Subterráneas y la Conservación de los Humedales: La Cuenca alta del Guadiana*; Coleto, M.C., Martínez Cortina, L., Llamas, M.R., Eds.; Fundación Marcelino Botín y Ediciones Mundi-Prensa: Madrid, Spain, 2003; pp. 3–68.

7. Castaño, S.; de la Losa, A.; Martínez-Santos, P.; Mediavilla, R.; Santisteban, J.I. Long-term effects of aquifer overdraft and recovery on groundwater quality in a Ramsar wetlands: Las Tablas de Daimiel National Park, Spain. *Hydrol. Process.* **2018**, *32*, 2863–2873. [CrossRef]

8. Martínez-Santos, P.; Castaño-Castaño, S.; Hernández-Espriú, A. Revisiting groundwater overdraft based on the experience of the Mancha Occidental Aquifer, Spain. *Hydrogeol. J.* **2018**, *26*, 1083–1097. [CrossRef]

9. Ahmed, K.; Shahid, S.; Demirel, M.C.; Nawaz, N.; Khan, N. The changing characteristics of groundwater sustainability in Pakistan from 2002 to 2016. *Hydrogeol. J.* **2019**, *27*, 2485–2496. [CrossRef]

10. Golodkovskaya, G.A.; Eliseyev, J.B. *Geological Media of Industrial Regions*; Moscow "Nedra" Publishing: Moscow, Russia, 1989; pp. 1–220.

11. Cuthbert, M.O.; Taylor, R.G.; Favreau, G.; Todd, M.C.; Shamsudduha, M.; Villholth, K.G.; MacDonald, A.M.; Scanlon, B.R.; Kotchoni, D.O.V.; Vouillamoz, J.M.; et al. Observed controls on resilience of groundwater to climate variability in sub-Saharan Africa. *Nature* **2019**, *572*, 230–234. [CrossRef] [PubMed]

12. Maurice, L.; Taylor, R.G.; Tindimugaya, C.; MacDonald, A.M.; Johnson, P.; Kaponda, A.; Owor, M.; Sanga, H.; Bonsor, H.C.; Darling, W.G.; et al. Characteristics of high-intensity groundwater abstractions from weathered crystalline bedrock aquifers in East Africa. *Hydrogeol. J.* **2019**, *27*, 459–474. [CrossRef]

13. Lorenzo-Lacruz, J.; García, J.; Morán-Tejeda, E. Groundwater level responses to precipitation variability in Mediterranean insular aquifers. *J. Hydrol.* **2017**, *552*, 516–531. [CrossRef]

14. Mazi, K.; Koussis, A.D.; Destouni, G. Intensively exploited Mediterranean aquifers: Resilience to seawater intrusion and proximity to critical thresholds. *Hydrol. Earth Syst. Sci.* **2014**, *18*, 1663–1677. [CrossRef]

15. Galassi, D.M.P.; Lombardo, P.; Fiasca, B.; Di Cioccio, A.; Di Lorenzo, T.; Petitta, M.; Di Carlo, P. Earthquakes trigger the loss of groundwater biodiversity. *Sci. Rep.* **2014**, *4*, 6273. [CrossRef]

16. Chinnasamy, P.; Maheshwari, B.; Prathapar, S.A. Adaptation of Standardised Precipitation Index for understanding watertable fluctuations and groundwater resilience in hard-rock areas of India. *Environ. Earth Sci.* **2018**, *77*, 562. [CrossRef]

17. De Eyto, E.; Jennings, E.; Ryder, E.; Sparber, K.; Dillane, M.; Dalton, C.; Poole, R. Response of a humic lake ecosystem to an extreme precipitation event: Physical, chemical, and biological implications. *Inland Waters* **2016**, *6*, 483–498. [CrossRef]

18. Everard, M.; Ahmed, S.; Gagnon, A.S.; Kumar, P.; Thomas, T.; Sinha, S.; Dixon, H.; Sarkar, S. Can nature-based solutions contribute to water security in Bhopal? *Sci. Total Environ.* **2020**, *723*, 138061. [CrossRef] [PubMed]

19. Somers, L.D.; McKenzie, J.M.; Mark, B.G.; Lagos, P.; Ng, G.-H.C.; Wickert, A.D.; Yarleque, C.; Baraër, M.; Silva, Y. Groundwater buffers decreasing glacier melt in an Andean watershed—But not forever. *Geophys. Res. Lett.* **2019**, *46*, 13016–13026. [CrossRef]

20. Gozzi, C.; Filzmoser, P.; Buccianti, A.; Vaselli, O.; Nisi, B. Statistical methods for the geochemical characterisation of surface waters: The case study of the Tiber River basin (Central Italy). *Comput. Geosci.* **2020**, *131*, 80–88. [CrossRef]

21. Hejazian, M.; Gurdak, J.J.; Swarzenski, P.; Odigie, K.O.; Storlazzi, C.D. Land-use change and managed recharge effects on the hydrogeochemistry of two contrasting atoll island aquifers, Roi-Namir island, Republic of the Marshall Islands. *Appl. Geochem.* **2017**, *80*, 58–71. [CrossRef]

22. MacDonald, A.M.; Bonsor, H.C.; Calow, R.C.; Taylor, R.G.; Lapworth, D.J.; Maurice, L.; Tucker, J.; Dochartaigh, O. *Groundwater Resilience to Climate Change in Africa*; British Geological Survey: Nottingham, UK, 2011.

23. Bouska, K.L.; Houser, J.N.; De Jager, N.R.; Van Appledorn, M.; Rogala, J.T. Applying concepts of general resilience to large river ecosystems: A case study from the Upper Mississippi and Illinois rivers. *Ecol. Indic.* **2019**, *101*, 1094–1110. [CrossRef]

24. Fraccascia, L.; Giannoccaro, I.; Albino, V. Resilience of complex systems: State of the art and directions for future reseaarch. Hindawi. *Complexity* **2018**, *2018*, 3421529. [CrossRef]

25. Helfgott, A. Operationalising systemic resilience. *Eur. J. Oper. Res.* **2018**, *268*, 852–864. [CrossRef]

26. Wurl, J.; Gámez, A.E.; Ivanova, A.; Imaz Lamadrid, M.A.; Hernández-Morales, P. Socio-hydrological resilience of an arid aquifer system, subject to changing climate and inadequate agricultural management: A case study from the Valley of Santo Domingo, Mexico. *J. Hydrol.* **2018**, *559*, 486–498. [CrossRef]

27. Lapworth, D.J.; MacDonald, A.M.; Tijani, M.N.; Darling, W.G.; Gooddy, D.C.; Bonsor, H.C.; Araguás-Araguás, L.J. Residence times of shallow groundwater in West Africa: Implications for hydrogeology and resilience to future changes in climate. *Hydrogeol. J.* **2013**, *21*, 673. [CrossRef]

28. Foster, S.; McDonald, A. The 'water security' dialogue: Why it needs to be better informed about groundwater. *Hydrogeol. J.* **2014**, *22*, 1489–1492. [CrossRef]

29. Fuchs, E.; Carroll, K.C. Quantifying groundwater resilience through conjunctive use for irrigated agriculture in a constrained aquifer system. *J. Hydrol.* **2018**, 747–759. [CrossRef]

30. Domenico, P.A. *Concepts and Models in Groundwater Hydrology*; International Series in the Earth & Planetary Sciences; McGraw Hill: New York, NY, USA, 1972; pp. 1–405.

31. Theis, C.V. The source of water derived from wells: Essential factors controlling the response of an aquifer to development. *Civil Eng.* **1940**, *10*, 277–280.

32. Sophocleous, M.A. Interactions between groundwater and surface water: The state of the science. *Hydrogeol. J.* **2002**, *10*, 52–67. [CrossRef]

33. Winter, T.C.; Harvey, J.W.; Franke, O.L.; Alley, W.M. Ground water and surface water: A single resource. *U.S. Geol. Surv. Circ.* **1998**, *1139*, 79.

34. Winter, T.C. Relation of streams, lakes, and wetlands to groundwater flow systems. *Hydrogeol. J.* **1999**, *7*, 28–45. [CrossRef]

35. Balleau, W.P. Water approximation and transfer in a general hydrogeologic system. *Nat. Resour. J.* **1988**, *29*, 269–291.

36. Rockström, J.; Steffen, W.; Noone, K.; Persson, Å.; Chapin, F.S.; Lambin, E.; Lenton, T.M.; Scheffer, M.; Folke, C.; Schellnhuber, H.; et al. Planetary Boundaries: Exploring the Safe Operating Space for Humanity. *Ecol. Soc.* **2009**, *14*, 32. Available online: http://www.ecologyandsociety.org/vol14/iss2/art32/ (accessed on 10 March 2020).

37. Dakos, V.; Carpenter, S.R.; van Nes, E.H.; Scheffer, M. Resilience indicators: Prospects and limitations for early warnings of regime shifts. *Phil. Trans. R. Soc. B* **2015**, *370*, 20130263. [CrossRef]

38. Linkov, I.; Bridges, T.; Creutzig, F.; Decker, J.; Fox-Lent, C.; Kröger, W.; Lambert, J.H.; Levermann, A.; Montreuil, B.; Nathwani, J.; et al. Changing the resilience paradigm. *Nat. Clim. Chang.* **2014**, *4*, 407–409. Available online: www.nature.com/natureclimatechange (accessed on 10 March 2020). [CrossRef]

39. Standish, R.J.; Hobbsa, R.J.; Mayfieldb, M.M.; Bestelmeyer, B.T.; Sudingd, K.N.; Battagliae, L.L.; Evinerf, V.; Hawkesg, C.V.; Tempertonh, V.M.; Cramera, V.A.; et al. Resilience in ecology: Abstraction, distraction, or where the action is? *Biol. Conserv.* **2014**, *177*, 43–51. [CrossRef]

40. Mitchell, T.; Harris, K. Resilience: A Risk Management Approach. Overseas Development Institute (ODI). 2012. Available online: https://www.odi.org/sites/odi.org.uk/files/odi-assets/publications-opinion-files/7552.pdf (accessed on 10 March 2020).

41. Miller, F.; Osbahr, H.; Boyd, E.; Thomalla, F.; Bharwani, S.; Ziervogel, G.; Walker, B.; Birkmann, J.; van der Leeuw, S.; Rockström, J.; et al. Resilience and vulnerability: Complementary or conflicting concepts? *Ecol. Soc.* **2010**, *15*, 11. Available online: http://www.ecologyandsociety.org/vol15/iss3/art11/ (accessed on 10 March 2020). [CrossRef]

42. Holling, C. Resilience and stability of ecological systems. *Annu. Rev. Ecol. Syst.* **1973**, *4*, 1–23. [CrossRef]
43. Lade, S.J.; Peterson, G.D. Comments on Resilience of Complex Systems: State of the Art and Directions for Future Research. Hindawi. *Complexity* **2018**, *2019*, 6343545. [CrossRef]
44. Gallopín, G.C. Linkages between vulnerability, resilience, and adaptive capacity. *Glob. Environ. Chang.* **2006**, *16*, 293–303. [CrossRef]
45. Walker, B.H.; Carpenter, S.R.; Rockström, J.; Crépin, A.S.; Peterson, G.D. Drivers, "slow" variables, "fast" variables, shocks, and resilience. *Ecol. Soc.* **2012**, *17*, 30. [CrossRef]
46. Folke, C.; Carpenter, S.; Walker, B.; Scheffer, M.; Elmqvist, T.; Gunderson, L.; Holling, C.S. Regime shifts, resilience, and biodiversity in ecosystem management. *Annu. Rev. Ecol. Evol. Syst.* **2004**, *35*, 557–581. [CrossRef]
47. Folke, C. Resilience: The emergence of a perspective for social-ecological systems analyses. *Glob. Environ. Chang.* **2006**, *16*, 253–267. [CrossRef]
48. Walker, B.; Holling, C.S.; Carpenter, S.R.; Kinzig, A. Resilience, adaptability and transformability in social-ecological systems. *Ecol. Soc.* **2004**, *9*, 5. [CrossRef]
49. Holling, C.S. The resilience of terrestrial ecosystems: Local surprise and global change. In *Sustainable Development of the Biosphere*; Clark, W.C., Munn, R.E., Eds.; Cambridge University Press: Cambridge, UK, 1986; pp. 292–317.
50. Gunderson, L.H. Ecological resilience—In theory and application. *Annu. Rev. Ecol. Syst.* **2000**, *31*, 425–439. [CrossRef]
51. Sundstrom, S.M.; Allen, C.R.; Barichievy, C. Species, Functional Groups, and Thresholds in Ecological Resilience. *Conserv. Biol.* **2011**, *26*, 305–314. [CrossRef]
52. Allen, C.R.; Angeler, D.G.; Garmestani, A.S.; Gunderson, L.H.; Holling, C.S. Panarchy: Theory and application. *Ecosystems* **2014**, *17*, 578–589. [CrossRef]
53. Eason, T.; Garmestani, A.S.; Stow, C.A.; Rojo, C.; Alvarez-Cobelas, M.; Cabezas, H. Managing for resilience: An information theory-based approach to assessing ecosystems. *J. Appl. Ecol.* **2016**, *53*, 656–665. [CrossRef]
54. Bertalanffy, L.V. *General System Theory*; Braziller: New York, NY, USA, 1968; pp. 1–289.
55. Scheffer, M.; Bascompte, J.; Brock, W.A.; Brovkin, V.; Carpenter, S.R.; Dakos, V.; Held, H.; van Nes, E.H.; Rietkerk, M.; Sugihara, G. Early-warning signals for critical transitions. *Nat. Rev.* **2009**, *461*. [CrossRef]
56. Clements, C.F.; Ozgul, A. Indicators of transitions in biological systems. *Ecol. Lett.* **2018**, *21*, 905–919. [CrossRef] [PubMed]
57. Angeler, D.G.; Fried-Petersen, H.B.; Allen, C.R.; Garmestani, A.; Twidwell, D.; Chuang, W.-C.; Donovan, V.M.; Eason, T.; Roberts, C.P.; Sundstrom, S.M.; et al. Adaptive capacity in ecosystems. In *Advances in Ecological Research*; Bohan, D.A., Dumbrell, A.J., Eds.; Academic Press: Cambridge, MA, USA, 2019; Volume 60, pp. 2–190. [CrossRef]
58. López Gutiérrez, J.; Ruiz Hernández, J.M.; y García de la Noceda, C. Caracterización Adicional de las Masas de Agua Subterránea en riesgo de no cumplir los objetivos medio ambientales en 2015. In *Las Aguas Subterráneas en la Planificación Hidrogeológica*; Fernández Ruiz, L., Ed.; Instituto Geológico y Minero de España: Madrid, Spain, 2012; pp. 1–481.
59. Krause, S.; Lewandowski, J.; Grimm, N.B.; Hannah, D.M.; Pinay, G.; McDonald, K.; Martı, E.; Argerich, A.; Pfister, L.; Klaus, J.; et al. Ecohydrological interfaces as hot spots of ecosystem processes. *Water Resour. Res.* **2017**, *53*, 6359–6376. [CrossRef]
60. Folke, C.; Carpenter, S.R.; Elmqvist, T.; Gunderson, L.; Holling, C.S.; Walker, B. Resilience and sustainable development: Building adaptive capacity in a world of transformations. *Ambio* **2002**, *31*, 437–440. [CrossRef]
61. Thomas, B.F.; Caineta, J.; Nanteza, J. Global assessment of groundwater sustainability based on storage anomalies. *Geophys. Res. Lett.* **2017**, *44*, 11445–11455. [CrossRef]
62. Scheffer, M.; Carperter, S.; Foley, J.A.; Folke, C.; Walker, B. Catastrophic shifts in ecosystems. *Nature* **2001**, *413*, 591–596. [CrossRef]
63. Suding, K.N.; Hobbs, R.J. Threshold models in restoration and conservation: A developing framework. *Trends Ecol. Evol.* **2009**, *24*, 271–279. [CrossRef]
64. Mediavilla, R. *Las Tablas de Daimiel: Agua y Sedimentos*; Mediavilla, R., Ed.; Instituto Geológico y Minero de España (Medio Ambiente; 14): Madrid, Spain, 2012; p. 285.

65. Ruíz-Zapata, B.; Gil-García, M.J.; de Bustamante, I. Paleoenvironmental reconstruction of Las Tablas de Daimiel and its evolution during the Quaternary period. In *Ecology of Threatened Semi-Arid Wetlands: Long-Term Research in Las Tablas de Daimiel*; Series Wetlands: Ecology, Conservation and Management; Sánchez-Carrillo, S., Angeler, D.G., Eds.; Springer: Dordrecht, The Netherlands, 2010; Volume 2, pp. 23–44.

66. De la Losa, A.; Aguilera, H.; Jiménez-Hernández, M.E.; Castaño, S.; Moreno, L. Hidrología e hidroquímica. In *Las Tablas de Daimiel: Agua y Sedimentos*; Mediavilla, R., Ed.; Instituto Geológico y Minero de España (Medio Ambiente, 14): Madrid, Spain, 2012; pp. 87–124.

67. Cirujano, S. Bentos vegetal, flora y vegetación superior. In *Las Tablas de Daimiel. Ecología Acuática y Sociedad*; Alvarez-Cobelas, M., Cirujano, S., Eds.; Organismo Autónomo Parques Nacionales, Ministerio Español de Medio Ambiente: Madrid, Spain, 1996; pp. 129–139.

68. Sánchez-Carrillo, S.; Álvarez-Cobelas, M.; Cirujano, S.; Carrasco-Redondo, M.; Díaz-Cambronero, A. Long-term observations as a key tool for the assessment of environmental changes in Las Tablas de Daimiel: LTER-Daimiel. *Ecosistemas* **2016**, *25*, 04–08. [CrossRef]

69. Aguilera, H.; Merino, L. Data on chemical composition of soil and water in the semiarid wetland of Las Tablas de Daimiel National Park (Spain) during a drought period. *Data Brief* **2018**, *19*, 2481–2486. [CrossRef] [PubMed]

70. Moreno, L.; Jiménez, M.E.; Aguilera, H.; Jiménez, P.; De la Losa, A. The 2009 Smouldering Peat Fire in Las Tablas de Daimiel National Park (Spain). *Fire Technol.* **2011**, *47*, 519–538. [CrossRef]

71. Aguilera, H.; Castaño, S.; Moreno, L.; Jiménez-Hernández, M.E.; De la Losa, A. Model of hydrological behaviour of the anthropized semiarid wetland of Las Tablas de Daimiel National Park (Spain) based on surface water/groundwater interactions. *Hydrogeol. J.* **2013**, *21*, 623–641. [CrossRef]

72. García, M. Hidrogeología de las Tablas de Daimiel y de los Ojos del Guadiana. Bases Hidrogeológicas Para una Clasificación Funcional de Humedales Ribereños. Ph.D. Thesis, Universidad Complutense de Madrid, Madrid, Spain, 1996. Available online: http://eprints.ucm.es/3104/ (accessed on 10 March 2020).

73. Rottler, E.; Francke, T.; Bürger, G.; Bronstert, A. Long-term changes in central European river discharge for 1869–2016: Impact of changing snow covers, reservoir constructions and an intensified hydrological cycle. *Hydrol. Earth Syst. Sci.* **2020**, *24*, 1721–1740. [CrossRef]

74. Whittemore, D.O.; Butler, J.J.; Wilson, B.B. Assessing the major drivers of water-level declines: New insights into the future of heavily stressed aquifers. *Hydrol. Sci. J.* **2016**, *61*, 134–145. [CrossRef]

75. UNESCO-IHP. *Management and Protection of Mediterranean Groundwater-Related Coastal Wetlands and Their Services*; United Nations Educational, Scientific and Cultural Organization: Paris, France, 2019.

76. O'Neill, R.V.; Johnson, A.R.; King, A.W. A hierarchical framework for the analysis of scale. *Landsc. Ecol.* **1989**, *3*, 193–205. [CrossRef]

77. Callahan, J.T. Long-term ecological research—An international perspective. In *Long-Term Ecological Research in the United States: A Federal Perspective*; Risser, P.G., Ed.; Willey: Chichester, UK; Reino Unido, UK; Hoboken, NJ, USA, 1991; pp. 9–21.

78. Chapin, F.S., III; Peterson, G.; Berkes, F.; Callaghan, T.V.; Angelstam, P.; Apps, M.; Beier, C.; Bergeron, Y.; Crépin, A.S.; Danell, K.; et al. Resilience and vulnerability of northern regions to social and environmental change. *Ambio* **2004**, *33*, 344–349. [CrossRef]

79. Martínez-Valderrama, J.; Ibáñez, J.; Alcalá, F.J.; Domínguez, A.; Yassin, M.; Puigdefábregas, J. The use of a hydrological-economic model to assess sustainability in groundwater-dependent agriculture in drylands. *J. Hydrol.* **2011**, *402*, 80–91. [CrossRef]

80. Alcalá, F.J.; Martínez-Valderrama, J.; Robles-Marín, P.; Guerrera, F.; Martín-Martín, M.; Raffaelli, G.; Tejera de León, J.; Asebriy, L. A hydrological-economic model for sustainable groundwater use in sparse-data drylands: Application to the Amtoudi Oasis in southern Morocco, northern Sahara. *Sci. Total Environ.* **2015**, *537*, 309–322. [CrossRef]

 *water*

Article

Climate-Dependent Groundwater Discharge on Semi-Arid Inland Ephemeral Wetlands: Lessons from Holocene Sediments of Lagunas Reales in Central Spain

Rosa Mediavilla [1],*, Juan I. Santisteban [2], Ignacio López-Cilla [1], Luis Galán de Frutos [1] and África de la Hera-Portillo [1]

[1] Geological Survey of Spain (IGME), C/Ríos Rosas 23, 28003 Madrid, Spain; i.lopezcilla@gmail.com (I.L.-C.); l.galan@igme.es (L.G.d.F.); a.delahera@igme.es (Á.d.l.H.-P.)

[2] Department of Geodynamics, Stratigraphy and Paleontology, Faculty of Geological Sciences, Complutense University of Madrid, C/José Antonio Novais, 12, 28040 Madrid, Spain; j.i.santisteban@geo.ucm.es

* Correspondence: r.mediavilla@igme.es

Received: 5 June 2020; Accepted: 1 July 2020; Published: 4 July 2020

Abstract: Wetlands are environments whose water balance is highly sensitive to climate change and human action. This sensitivity has allowed us to explore the relationships between surface water and groundwater in the long term as their sediments record all these changes and go beyond the instrumental/observational period. The Lagunas Reales, in central Spain, is a semi-arid inland wetland endangered by both climate and human activity. The reconstruction of the hydroclimate and water levels from sedimentary facies, as well as the changes in the position of the surface water and groundwater via the record of their geochemical fingerprint in the sediments, has allowed us to establish a conceptual model for the response of the hydrological system (surface water and groundwater) to climate. Arid periods are characterized by low levels of the deeper saline groundwater and by a greater influence of the surface freshwater. A positive water balance during wet periods allows the discharge of the deeper saline groundwater into the wetland, causing an increase in salinity. These results contrast with the classical model where salinity increases were related to greater evaporation rates and this opens up a new way of understanding the evolution of the hydrology of wetlands and their resilience to natural and anthropogenic changes.

Keywords: wetlands; paleo-groundwater; climate; sedimentary facies; geochemistry; Holocene; Spain

---

1. Introduction

Wetlands (enclosing coastal, fluvial and lake systems) are a valuable resource for mankind as they play a key role in climate regulation, house more than 40% of the living species of the planet and are an important element of the water cycle [1–3]. However, they are one of the most threatened ecosystems and have lost more than 50% of their surface during the last century due to human activities [4,5].

The Iberian Peninsula is one of the European regions with the greatest diversity of wetlands, particularly given the number of semi-arid to arid inland wetlands [6,7]. From the 1950s to 1960s, nearly 40% of the Spanish wetlands were drained for farming [8] and, in the following decades, the drained wetland surface increased due to the abstraction of their feeding groundwater. To evaluate the extent of the effects of human activity on wetlands and their resilience to these actions, it is necessary to analyse the long-term evolution of the water–wetland linkage and to compare the behaviour of the system under natural and human-influenced conditions.

To achieve these goals, most common strategies are based on the analysis of the water budget (water input vs. output in relation to the wetland's flooded surface) and the comparison of "natural" vs. "human-controlled" wetlands. These approaches rely on instrumental timeseries data (piezometric levels, meteorological data, gauge station data and remote sensing) that are usually too short or use remote areas with low human intervention, so a comparison between the "natural" and "human-controlled" stages has not been possible in the same wetland.

Nevertheless, there are some long records of these wetlands, which have been studied from their sediments. These studies have focused on the reconstruction of climate and hydrological evolution both in saline lakes in southern [9–13] and northern Spain [14–17], as well as in other non-saline lakes [18,19]. These studies, amongst others, have identified alternating drier and wetter periods (Roman Period, Early Middle Age, Medieval Climate Anomaly, Little Ice Age and Industrial Era) that have conditioned the hydrological evolution of the lakes. There are also a considerable number of studies about the record of the environmental changes caused by human intervention in wetlands [20–23].

However, despite it being assumed that a lake/wetland's water levels depend on changes in groundwater levels, there have been few studies that have investigated this link using the sedimentary record. Most of them have dealt with surface water and groundwater level relationships via water balances [24–26] or climate vs. water level/quality [26,27], but they have always considered overall water composition and salinity as a whole without taking into account the sources of the water (deep long residence time water, surface aquifers, etc.). In this sense, the use of groundwater proxies, such as As, Cl or Br, is restricted to their use as a source of information about the origin of the waters [28–30] but not as a tool to reconstruct the changes in the groundwater in the past centuries.

Thus, as concluded in the previous paragraphs, in order to improve our understanding about wetland–groundwater interactions, it is necessary to integrate both sources of knowledge (recent hydrogeological and long-term records of water balances). To achieve this goal, this paper presents a reconstruction of the water balance and its composition from the sedimentary facies and geochemical records of a semi-arid ephemeral wetland in central Spain, as well as of the linkages amongst the surface water–groundwater levels as a result of the mixing of waters with different origins and under different climate settings. Previous sedimentary models [24–27] used available aquifer conceptualization to reconstruct the hydrogeological evolution from sediments. In this example, we use the sedimentary record to understand the aquifer behaviour before reconstructing its evolution.

2. Study Area

2.1. Location

The Lagunas Reales wetland is a Protected Natural Area included in the Castille-León List (1993) (codes: VA-3 and VA-4), Special Protection Area for Birds (ZEPA: ES0000204, Tierra de Campiñas) and Location of Community Interest (LIC: ES4180147, Humedales de los Arenales). It is located in the Northern Spanish Meseta, a few kilometres south from Medina del Campo, near the confluence of the Zapardiel River and its tributary, Arroyo Simplón (Figure 1a,b).

Geologically, they are near the southern border of the Cainozoic Duero Basin and they lie upon Cainozoic fluvial arkosic sediments (partially cemented gravels and sands with clayey and carbonate crust levels) of Miocene age [31]. They belong to the Medina del Campo groundwater body [32].

The wetland is composed of two ephemeral endorheic ponds, connected by two small channels, lying above 720 m a.s.l. (Figure 1c). Their maximum depth is 0.5 m and they cover an area of 5.8 ha (the southern pond) and 5.0 ha (the northern one, which is the focus of this study). Currently, the ponds are dry and surrounded by pine tree forests and croplands.

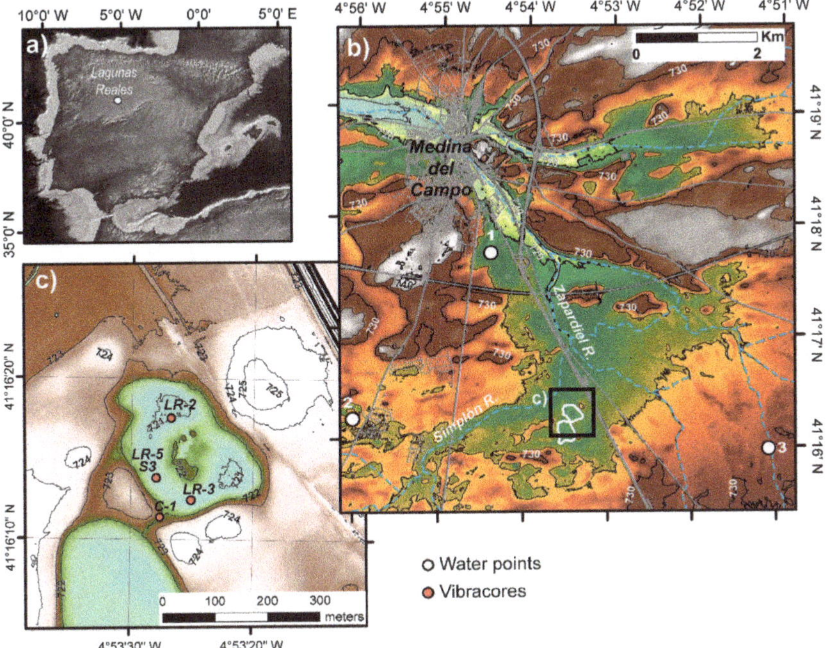

Figure 1. Location of the study area. (**a**) Location of the Lagunas Reales in the Iberian Peninsula. (**b**) Coloured digital elevation model showing the location of the wetland in relation to the Zapardiel and Simplón rivers. Black lines: 5 m contour lines. Water points: (1) 1517-4-002, (2) Balneario de Las Salinas and (3) N-64. (**c**) Location of studied cores (red dots) in the northern pond. Black lines: 1 m contour lines.

2.2. Climate, Hydrology and Historical Evolution of the Wetland

Most of the knowledge related to the Lagunas Reales wetland was obtained from recent timeseries compiled in official databases about meteorology (Spanish Meteorological Agency, AEMET), hydrology (Water Points Database from the Geological Survey of Spain, IGME, databases about groundwater levels and water composition of the Duero Basin Water Authority, CHD, and topographical and remote sensing information from the Geographical Institute of Spain, IGN) and land-use and human activity data (Statistical Institute of Spain, INE, and Chamber of Agriculture of Medina del Campo).

The area shows a Bsk (cold steppe arid) climate, accordingly to the Köppen–Geiger classification, with warm summers, cold winters and dry summers with rainfall seasons in autumn–early winter and spring. For the 1981–2020 period, the mean annual rainfall was 351 mm, mean annual temperature 12.9 °C and potential evapotranspiration (PET) 730 mm [33].

The longest rainfall time series, covering the 1851–2019 period, corresponds to the Valladolid weather station, which is located 40 km to the north and shows a good correlation with the data from the Medina del Campo station, which covers the 1932–2000 period (Figure 2a). Both series show an overall rising trend in annual rainfall.

The population of Medina del Campo and percentage of irrigated vs. non-irrigated cropland time series for the 20th and 21st centuries show a rising trend (Figure 2b). Despite the increase in population being constant for the 20th century, it speeds up from the 1960s to 1980s, coeval to the greatest increase in irrigated surface that changed from 22 ha in 1881 to 261 ha in 1960, and then to 1600 ha in 1986. After that moment, both series slowed down and decreased from 2010 (the global economic crisis) until the present day. Currently, irrigated crops represent nearly 30% of the total cultivated area and the population is above 20,000 inhabitants.

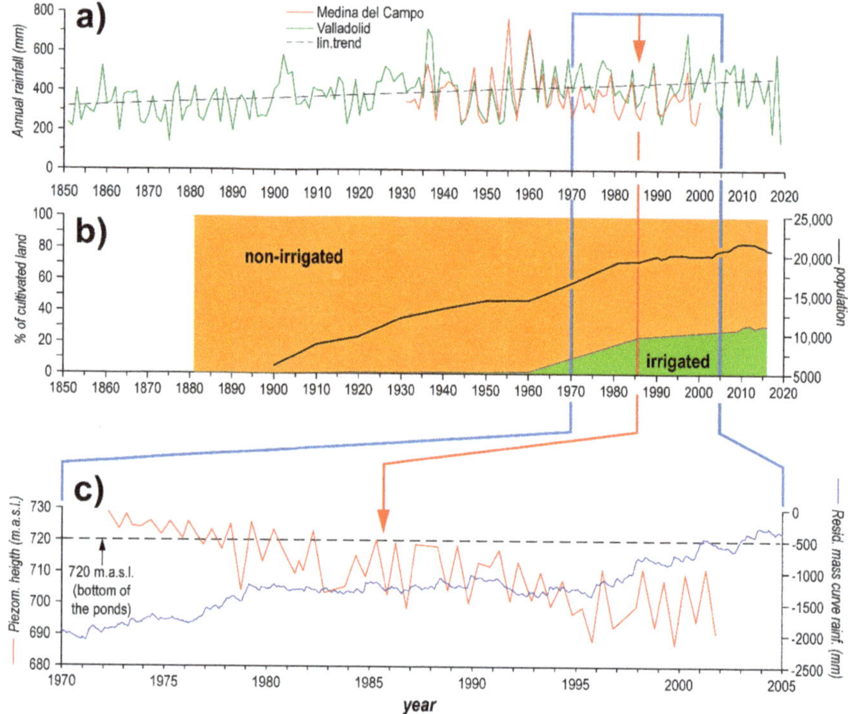

Figure 2. Climate, population, land-use and water-level evolution. (**a**) Annual rainfall series for the Valladolid and Medina del Campo weather stations (AEMET); (**b**) population (INE) and percentage of irrigated vs. non-irrigated cultivated area (Chamber of Agriculture of Medina del Campo); (**c**) piezometric height for water point 1517-4-0002 (IGME) (base level 730 m a.s.l., depth: 140 m) and the residual mass curve for rainfall for the 1971–2001 period. This period is bracketed by a blue line in panels (**a**,**b**). The fall in1985 is marked by a red arrow in the three panels.

Due to the climate of the region (with a PET greater than the annual rainfall), water supply for urban and farming uses depends on groundwater and water demand has increased continuously.

When the piezometric height from water point 1517-4-0002 (near Medina del Campo, indicated as 1 in Figure 1b) is compared with the residual mass curve for rainfall it is evident that despite the increase in rainfall for the 1971–2001 period, the piezometric height shows a lowering trend for the same period (Figure 2c), which can be attributed to human exploitation.

A key date in this trend is 1985, when the piezometric level fell below 720 m a.s.l., the lowermost point of the Lagunas Reales (Figures 1c and 2c) [34]. Until this date, the wetland was fed by rainfall, surface runoff and groundwater, and it was a discharge area for the multilayer aquifer [35]. From the fall of 1985 onwards, the wetland has only been fed by surface runoff and rainfall and it only shows a fully developed water body in very rainy years (1991, 1998, 2001 and 2018), when it became a recharge area for the aquifer.

Data about the behaviour of the Lagunas Reales under "natural" conditions (prior to the beginning of the human over-exploitation of the groundwater in the 1960s) are aerial photographs that point to an ephemeral water body with seasonal and inter-annual fluctuations. Thus, in July of 1946, a dry year, the ponds were dry whilst in June of 1956, a wet year, they were flooded, stressing the role of climate on both surface and groundwater supply to the wetland.

There is no information about the present chemical composition of the water as they have been dry for most of the past years. Salinity of the regional aquifer has been considered as related to recharge waters [36]. The Lagunas Reales and adjacent wetlands were, before 1985, a discharge area of sub-regional groundwater fluxes, supplying mineralized waters responsible for the saline efflorescences found in the wetland and the surroundings [37–40]. Waters from the N-64 water point (indicated as 2 in Figure 1b) are rich in Cl^--Na^+; those from Balneario de Las Salinas (indicated as 3 in Figure 1b) and N-61 (10 km to the south) are rich in HCO_3^--Cl^--Na^+-Ca^{2+} whilst those from PZ 02.17.25 (15 km to the south) are rich in HCO_3^--Na^+-Ca^{2+} (Table 1). There are other documented examples of brackish wetlands resulting from the mixing of saline groundwater and surface water of diverse composition in the proximity of the study area, such as the Lagunas de Villafáfila [41,42] and the Coca–Olmedo Pond Complex [43]. Consequently, we can assume that the water composition of the wetland resulted from a variable mixture of mineralized groundwater and low-mineralized or non-mineralized precipitation and surface water.

Table 1. Groundwater chemical analysis for the water points near the Lagunas Reales (IGME, CHD). N-64 and Balneario Las Salinas correspond to points 1 and 3 in Figure 1, respectively. N-61 and PZ 02.17.25 are out of the mapped area.

	N-64	Balneario Las Salinas	N-61	PZ 02.17.25
Depth (m)	61	20	71	120
Sample date	31 January 2000	18 October 2005	03 February 2000	24 May 2007
pH	9.56	8.51	7.30	7.82
Conductivity (20 °C) µS/cm	918.00	1425.00	300.00	490.00
Alkalinity mg/L $CaCO_3$	124.90	209.70	62.40	173.72
Cl^- mg/L	175.30	266.20	43.80	57.35
CO_3^{2-} mg/L	98.00			<5.00
HCO_3^- mg/L	26.90	137.30	62.40	211.81
PO_4^{3-} mg/L	0.65	<0.05	0.12	0.73
Ca^{2+} mg/L	1.90	116.70	25.80	48.42
Mg^{2+} mg/L	1.04	26.80	2.90	5.86
Na^+ mg/L	216.70	163.30	39.40	46.21
K^+ mg/L	3.40	5.90	2.30	1.33
NH_4^+ mg/L	0.07	0.09	0.04	<0.04
NO_2^- mg/L	0.42			<0.04
NO_3^- mg/L	26.81	10.10	0.00	8.30
Fe mg/L	0.56	<0.01	0.09	<0.05
Mn mg/L	0.03	0.32	0.00	<0.02
Zn mg/L	0.05		0.00	
SiO_2 mg/L		18.19		23.43
SO_4^{2-} mg/L		139.50	23.00	14.11
Sr mg/L		1.97		

3. Materials and Methods

In January 2018, we carried a coring campaign at the Lagunas Reales. Nine cores with a portable Ejkelkamp vibracoring system (total recovered length: 14.1 m; diameter: 5 cm) and 4 percussion cores (total recovered length: 17.82 m, diameter: 6.2 cm) were recovered. The cores were divided in two in the IGME laboratories and a half was stored as an archive, whilst the other half was used for description, analyses and sampling. All of them were photographed, their sedimentary features were described, and stratigraphic logs were elaborated correcting the depths to remove the effect of compaction due to drilling. In this paper, only cores C1, LR2, LR3 and LR5, recovered in the northern pond (Figure 1c), are presented.

Non-destructive analyses were run on the cores, at the IGME laboratories, including:

- XRF scanning with a GEOTEK XRF core scanner in a He purged atmosphere with an illumination window of 15 mm (cross-core slit width) × 10 mm (down-core resolution). Two runs, with a 30 s count time exposure, were performed for 10 kV and 40 kV (detecting from Mg to U). XRF spectra were processed with bAxil. Element intensities are represented in peak area.

- Lightness (L*) analyses were performed with 1 cm down core resolution using a Konica Minolta 700-d spectrophotometer integrated in the GEOTEK XRF core scanner. Lightness values range from 0 (black) to 100 (white).
- Geophysical properties (P-wave velocity, gamma density, non-contact resistivity and magnetic susceptibility) were analysed with a 1 cm down core resolution with a GEOTEK Multi-Sensor Core Logger (MSCL-GEOTEK). In this paper, gamma density was used because of its correlation with other logs. Gamma density data were obtained by the attenuation of a collimated gamma ray beam (5 mm diameter) emitted from a ^{137}Cs sealed source, passing through the core.
- Core colour scans (high resolution images with a down core resolution 50 µm) were obtained using a Geoscan IV coupled to the MSCL GEOTEK.

Destructive sampling was carried out in cores LR2, LR3 and LR5 to characterize the sedimentary facies at the IGME laboratories:

- Mineralogical analysis by X-ray diffractometry (PTE-RX-04) for the bulk sample and <2 m fraction. These analyses were used to check the sources of the chemical elements obtained from the geochemical analyses.
- Geochemical analysis of the major oxides and trace elements by X-ray fluorescence and atomic absorption spectroscopy (XRF and AAS). The results were used to check the validity of the non-destructive high-resolution XRF scanning.
- C (organic, inorganic and total) and S by elemental analyser (ELTRA). S data were used to check the results of the XRF core scanning. C values gave an estimate of organic matter and carbonate content and they can be compared to other results from non-destructive techniques (XRF core scanning and L*).

Selected samples were chosen for ^{14}C (AMS) dating of cores LR2, LR3 and LR5. Bulk samples of clays with organic matter were analysed at the Gliwice Absolute Dating Methods Centre (GADAM Centre, Gliwice, Poland). Dates were calibrated by the OxCal 4.2.4 calibration program [44] using the atmospheric IntCal13 curve [45] and expressed as Common Era (CE) and Before Common Era (BCE) calendar dates (Table 2).

Table 2. Radiocarbon ages and calibration. A sample name is composed of the name of the core-number of the sample + the depth of the sample.

Lab. Number	Sample Name	^{14}C Age Year (BP)	1σ Range cal. Age (CE/BCE)	2σ Range cal. Age (CE/BCE)
GdA-5576	LR2-1/91–95 cm	945 ± 30	1082–1128 CE	1025–1157 CE
GdA-5577	LR2-4/192–196 cm	2315 ± 30	403–376 BCE	412–356 BCE
GdA-5578	LR3-2/145.5–147.5 cm	1170 ± 30	855–894 CE	772–904 CE
GdA-5579	LR5-1/22.5-26.5 cm	135 ± 20	1834–1878 CE	1799–1890 CE
GdA-5580	LR5-3/133–136 cm	1030 ± 30	989–1023 CE	962–1042 CE
GdA-5581	LR5-5/161–165 cm	1585 ± 35	486–535 CE	400–550 CE

Geochemical data derived from the XRF core scanning were represented and statistically processed as single elements, but due to the diverse sources for several elements (chemical, terrigenous or organic matter-related) several ratios were elaborated. Normalization with Al was used to take account of the terrigenous fraction [46], and the Ca/S ratio was used to assess the Ca amount not related to gypsum [47]. Ratios related to redox state (Mn/Fe, [48,49]) or siliciclastic grains vs. the matrix (Si/Al, [47]) were also used. Due to the complex compositional relationships (elements with diverse origins or different datasets depending on the technique), principal component analysis (PCA) was carried out on the raw dataset, augmented with the lightness values, to explore the relationships amongst the elements. The analysis was carried in the R environment by using the routine "prcomp" [50] with the options "scale" and "center" set to TRUE to normalize the values. For this PCA, all the wetland deposits were analysed by cores and as a whole dataset, resulting in no differences amongst the cores.

4. Results

4.1. Sediments

The description of the sediments filling the Lagunas Reales and their Miocene substratum can only be made from the recovered cores.

4.1.1. Substratum of the Wetland

The Miocene arkoses are found in cores C1 and LR3 (Figure 3) and are made up of fine gravel and sand. Gravel appears in 2 to 5 cm-thick layers composed of quartz and quartzite clasts, with clast sizes from 2 to 5 mm (maximum 1.5 cm), and a sandy to silty matrix. Sandy levels are made up of medium sand with a silty to clayey matrix and calcite nodules are common. Both lithologies show the same mineralogical composition for the sandy, silty and clayey fractions: quartz, microcline, Ca-bearing albite and phyllosilicates (illite and smectites and minor quantities of muscovite, kaolinite and chlorite).

The sediments show an ochre colour that progressively changes to green (through a mottled area) at the boundary with the wetland deposits. These changes in colour are indicative of alterations in reduction and oxidation of the sediment due to wetting–drying phases in a poorly drained environment [51,52]. This hydromorphic feature plus the presence of clay illuviation and mm-sized crystals of lensoidal gypsum, sometimes related to root traces (core LR3, Figure 3), are typical of gleyed protosols [53] developed in relation to saline groundwater and the beginning of the development of the pond. The presence of such paleosoil levels at different heights in cores C1 and LR3 (Figure 3) points to the multiple pedogenic episodes.

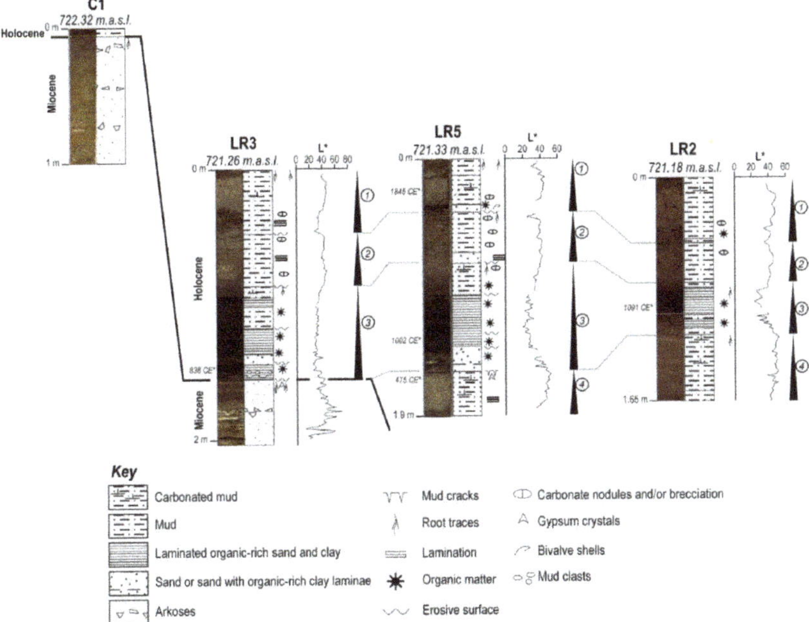

Figure 3. Stratigraphic sections for cores C1, LR2, LR3 and LR5 placed at their respective heights. Lightness (L*) is represented as indicative of the darker, organic-rich levels.

4.1.2. Sedimentary Fill of the Ponds

The sedimentary infill of the Lagunas Reales reaches 3.1 m in core S3 (Figure 1c) and, using the lineal age model developed for core LR5 (Figures 1c and 4), it can be assumed that the ponds are at least 3000 years old. The record for the last millennium, the focus of this paper, is composed of siliciclastic sediments with variable content in organic matter and/or carbonate that are arranged in four fining-upwards sequences (Figure 3). The described sequences are laterally homogeneous so there are no changes amongst the cores except for the vertical direction.

- Sequence 1: It is represented in all the cores with thicknesses ranging from 0.39 to 0.47 m (Figure 3). According to the age model, this sequence covers from 1700 to 1800 CE until the present (Figure 4). Its lower boundary is an erosive surface and the upper one is the bottom of the present-day ponds and it shows pedogenic features (root traces, E horizon in the uppermost cm). They are grey to green silty sands and mud with a variable amount of sand-sized grains forming a fining-upwards sequence. Their mineralogy is composed of quartz, microcline, Ca-bearing albite, phyllosilicates (illite and smectite) and calcite, the latter mixed with the clay (increasing upwards) or forming nodules of up to 2 mm in size (downwards). There is sparse organic matter and some bivalve fragments can be found in the sandy layers. They show horizontal lamination, which can be diffuse where carbonate nodules and root traces (<2 mm in diameter) are present (mainly downwards), and erosive surfaces are common. This sequence records the infill of the ponds, from the initial flooding stage, when the siliciclastic sediment is supplied by surface runoff, until their present-day stage. The internal erosive surfaces document different runoff episodes, probably linked to heavy rainfall episodes that brought the sediment from the surrounding reliefs and flooded the ponds. Clay flocculation was the main process during the high-water stages while calcite precipitation took place during the moments of greater evaporation. The presence of root traces and horizontal lamination point to shallow and calm waters whilst the presence of carbonate nodules is indicative of some pedogenic episodes linked to the temporary drying-out of the ponds [54,55]. The decreasing upwards intensity of the pedogenic processes and the increase in subaqueous precipitation of calcite point to a positive water budget trend and a more stable water body.
- Sequence 2: This sequence is represented in all the cores with thicknesses ranging from 0.29 to 0.39 m (Figure 3) and its age is between 1400 and 1500 CE and 1700 and 1800 CE (Figure 4). Its lower boundary is an erosive surface and its top is cut by Sequence 1. It is a fining-upwards sequence composed of green to dark green silty sand at the bottom, and poorly laminated to massive mud, with variable content in sand-size grains, dominating most of the sequence. Its mineralogical composition is similar to Sequence 1; however, the carbonate nodules (up to 1 cm in size) are more abundant, always related to root traces and increasing upwards, and there is sparse organic matter, mm-sized charcoal particles and shell fragments. Like Sequence 1, this sequence records from an initial flooding, linked to high energy events able to erode the bottom of the pond, until a final stage when dry periods were more frequent than the flooded ones. Each sedimentation episode records an initial flood, recorded by sands, followed by clay flocculation on the temporary water body and pedogenesis when the pond dried out. However, unlike Sequence 1, the increase upwards in pedogenic features points to longer and more frequent sub-aerial exposure episodes, which would result from an overall negative water budget.
- Sequence 3: The sequence thicknesses vary from 0.38 m (core LR2) up to 0.80 m (core LR5) (Figure 3). It is dated between ca. 525 and 675 CE and 1400 and 1500 CE (Figure 4). Its lower boundary is an erosive surface while its top is the erosive bottom surface of Sequence 2. It is a fining-upwards sequence made up of, from bottom to top, an alternation of gravel, mostly composed of mud clasts resulting from the erosion of the unit below (mean size: 3 mm; maximum size: 2 cm), and mud, changing upwards into inter-layered sands and mud, and ending with carbonated mud. The whole sequence shows well-defined lamination and pedogenic features (root traces and mm-sized carbonate nodules), restricted to the top of the sequence. The silicate composition is similar to the upper sequences, where calcite can reach up to 30% in the upper carbonated mud, and the more distinctive feature of this sequence is the higher organic matter content, in comparison to sequences 1, 2 and 4, which imprints darker tones to the sediment (dark green to dark grey). Like the previous ones, this fining-upwards sequence records the progressive infill of the pond, from an initial flood-dominated stage to a final desiccated stage via

successive flooding episodes. The main difference with the other sequences is that the greater amount of organic matter of vegetal origin and the minor development of pedogenic features (only to the top) imply wetter conditions, a stable water body and an overall positive water budget when compared to the other sequences.

- Sequence 4: This was only recorded in cores LR2 (thickness: 0.50 m) and LR5 (thickness: 0.33 m) (Figure 3). The lower boundary cannot be seen and the upper one is the erosive bottom surface of Sequence 3. According to the age model, this sequence is older than 525–675 CE (Figure 4). It is composed of carbonated mud to marl with variable amounts of sand and clay. The siliciclastic fraction is composed of phyllosilicates (illite, smectite, muscovite and chlorite), quartz, microcline and Ca-bearing albite. The calcite content ranges from 36% to 50%. The sediment shows a cream or pale grey colour and may contain mm-sized fragments of charcoal and shells. It is arranged in sets of parallel laminae bounded by erosive surfaces. Root traces (mm-size in diameter) are present throughout the sequence but increase to the top of the sequence, where mm-size carbonate nodules and lensoidal crystals of gypsum appear. Its genesis is like that of Sequence 2, but the amount of calcite and the presence of gypsum point to higher evaporation rates, and the ubiquity of rootlet traces are indicative of a very shallow water body.

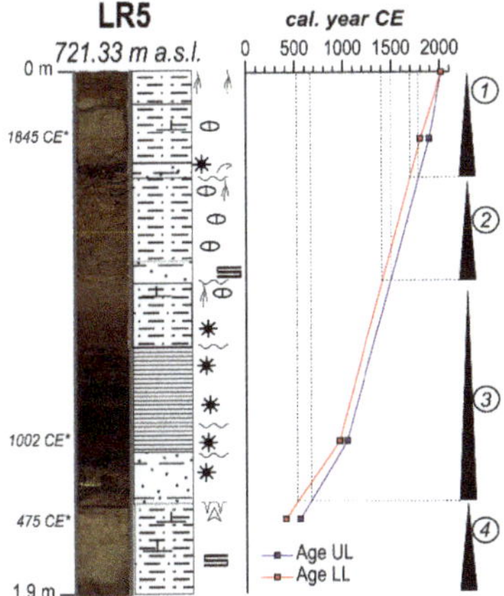

Figure 4. Age model for core LR5. Same key as for Figure 3. UL: upper age limit; LL: lower age limit.

4.2. Geochemical and Geophysical Record

PCA analysis on the raw geochemical dataset, augmented with the lightness (L*), reveals clusters of elements/parameters that may be genetically related. Those related to silicates (Si, K, Al, Ti and Fe), redox conditions (Mn) and saline waters (S, Sr, Br, Ca, As, Cl and Mg), as well as those linked to the organic matter (U, P and lightness) (Figure 5).

4.2.1. The Miocene/Holocene Boundary

These parameters are helpful to distinguish between the arkoses of the substratum and the Holocene sediments of the wetland. Despite being derived from the arkoses, some geochemical and geophysical features allow us to identify the boundary between both geological units (Figure 6a). Gamma ray density (GDR) shows a sharp boundary marked by a decrease in density in relation to the basal lag made of mud clasts. There is no clear boundary when the siliciclastic fraction is considered (Si and Al). However, sulphur (S) shows a noticeable change across the boundary. The presence of a hydromorphic paleosoil with gypsum at top of the arkoses was recorded with higher values of S that suddenly decrease at the bottom of the Holocene deposits. Ca also shows a sudden rise but not as marked as that of S. However, the most striking change is that of Sr, Cl and Br, showing much lower values in the Miocene arkoses than in the Holocene deposits, what is indicative that these sediments are not the source of such elements in the water.

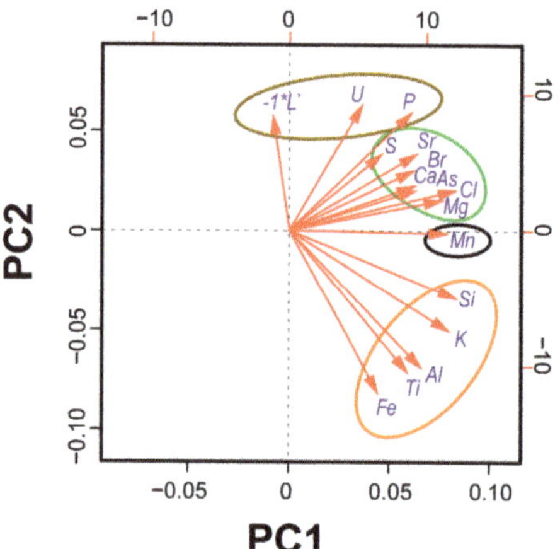

Figure 5. Biplot of the two principal components (PC) of the raw geochemical dataset. Elements with higher correlation coefficients (co-dependent) were cleaned and the lightness was expressed as −1* L* in order to stress its relationship to the organic matter. Clusters: brown: organic matter related elements; green: groundwater affine elements; black: redox sensitive element; orange: siliciclastic fraction.

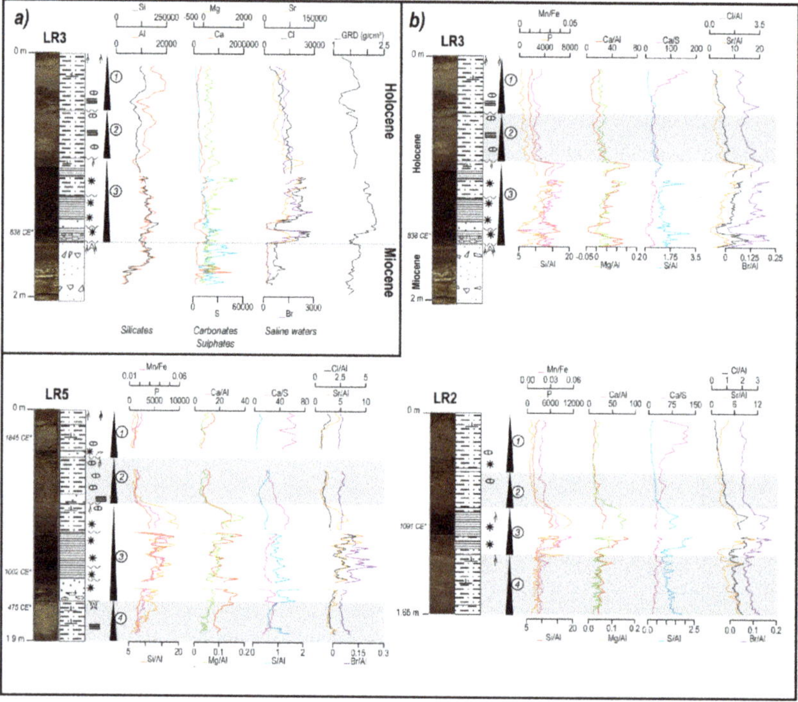

Figure 6. Geochemical logs for the studied cores, with the same key as for Figure 3. (**a**) Single element plots and gamma ray density (GRD). (**b**) Geochemical ratios and P log for the studied cores. Grey rectangles: sequences characterized by a greater amount of pedogenic features.

4.2.2. Geochemical Record of the Holocene Sequences

Due to the dominance of the siliciclastic fraction and the mixed origin for many of them, when plotted along the cores, they do not show clear variations to help in the correlation of the Holocene sediments. To filter the effect of the weight of the siliciclastic fraction, normalization with Al is used [46].

In addition, the Ca/S ratio allows us to evaluate the amount of Ca from silicates and carbonate against that included in gypsum; the Si/Al ratio accounts for the excess of Si in silicates, which can be correlated with an increase in quartz vs. other silicates and, secondarily, to a grainy/clayey matrix ratio [47]. Finally, the Mn/Fe ratio, indicative of redox conditions [49], and P, highly correlated to L* and organic carbon, are proxies for the organic matter. The described sequences can also be characterized by means of their geochemical features (Figure 6b). One fact that must be highlighted is that all the sequences are bounded by erosive surfaces (except for the top of Sequence 1), which implies that the sequences have incorporated material from the underlying sequences at the bottom and that they are truncated, showing a more or less incomplete record at the top.

- Sequence 1: Its lower boundary is pinpointed by an increase in Mn/Fe, Ca/Al, Mg/Al, S/Al, Sr/Al, Br/Al and Cl/Al in LR3, which can be related to the initial flooding stage of the sequence, and by a change in trend for P, Si/Al, Ca/S and Sr/Al for LR2, which may be interpreted as the result of the change from drying (Sequence 2) to wetting (Sequence 1) conditions. The uppermost centimetres of the sequence show a slight increase in Sr/Al, Cl/Al and S/Al coeval with a decrease in the Ca/S ratio that could be ascribed to a salinity increase related to the recent drying of the ponds. This sequence is characterized by low values of Si/Al, Mn/Fe and P, which are coherent with its mostly clayey composition and low organic content. The Ca/S ratio increases upwards despite the Si/Al, S/Al, Ca/Al, Mg/Al stand still and Cl/Al, Sr/Al and Br/Al remain the same or show a slow decrease. This implies an increase of Ca that is not due to silicate or gypsum minerals and is compatible with the described increase in calcite due to precipitation from a relatively stable water body.

- Sequence 2: Its lower boundary is marked by a noticeable decrease in the values of all the geochemical parameters in relation to Sequence 3. As compared to Sequences 1 and 3, Sequence 2 shows low values in siliciclastic, organic matter and redox proxies (Si/Al, Mg/Al, Ca/Al, P, Mn/Fe) and saline elements (S/Al, Sr/Al, Cl/Al and Br/Al), as well as in the fresh vs. saline waters ratio (Ca/S), which points to arid conditions (scarce surface water inputs, poorly developed vegetation and evaporative conditions).

- Sequence 3: Its bottom can only be observed in LR2 and LR5 and it is signalled by a noticeable increase, and change in trend, for all the proxies in LR2 and, in LR5, by a break in P and Mn/Fe, a rise in Ca/Al and S/Al, drop in Ca/S and change in trend for Sr/Al, Cl/Al and Br/Al. In relation to Sequences 2 and 4, Sequence 3 shows higher values for all the geochemical proxies which, in addition, shows an increasing upwards trend for the lower part of the sequence followed by progressive (LR2, LR5) or sudden (LR3) decrease of all the values, except for S/Al that shows a continuous decreasing upwards trend. The base of this sequence is characterized by anomalous low values in elements linked to the mineralogical composition (Si/Al, Mg/Al, Ca/Al, etc.), which is due to the incorporation of eroded material from the lower sequence in addition to the sediment supplied by surface water. It is worth mentioning that there is a good peak-to-peak direct correlation amongst all the proxies. The correlation amongst the proxies for surface inputs (Si/Al, P, Mn/Fe) and those related to the presence of saline water (Ca/Al, Mg/Al, Sr/Al, Cl/Al, Br/Al) point to a mixture of both sources but the decrease in S/Al and increase in Ca/S reveal a dilution, so the sediments become less saline higher up.

- Sequence 4: This sequence shows a decrease in all the proxies as compared to Sequence 3. P, Mn/Fe, Mg/Al, Ca/Al and Ca/S have a slightly increasing upwards trend, more evident for LR5 than for LR2, whilst S/Al, Br/Al, Cl/Al and Sr/Al show a decreasing trend. On the other hand, the Si/Al ratio is almost constant in both cores. As compared to Sequence 3, this sequence points to lower surface inputs and a decrease in the of supply saline elements, which is coherent with a drier period; however, in comparison to Sequence 2, the increasing upwards trend of the Ca/S ratio and other ratios seem to point to wetter conditions for that period.

5. Discussion

The presented record is ruled by very simple principles that allow its interpretation in terms of allocyclic or autocyclic forcings.

In our example, the principles controlling the sedimentation are the following:

(1) Clastic sedimentation is related to surface runoff and, therefore, to rainfall episodes.
(2) Chemical sedimentation takes place under the surface of a "stable" water body.
(3) Vegetation development is enhanced by favourable ecological conditions (water availability).
(4) Pedogenesis takes place where/when the water table is below the ground surface.
(5) The sequences cover from the initial flooding stage to the drying out of the pond.
(6) A sequence records a certain number of events (depositional or not) that took place during its recorded period.

On the basis of these principles we can consider the following:

(1) An increase in pedogenesis implies a longer total period without a water table above the ground surface.
(2) Chemical sediments suggest a relatively stable water table and warm conditions that increase evaporation and the concentration of solutes.
(3) An increase in clastic sedimentation implies more frequent rainfall episodes.
(4) An increase in organic matter indicates wetter conditions.
(5) The shallowing upwards sequence implies a destruction of accommodation space. This can be achieved by a lowering of the water table (drying out), the rising up of the bottom (silting of the pond) or a combination of the two.
(6) The sequences record the result of a combination of processes at different time scales and the resulting trend shows the long-term evolution of these processes. Thus, the sequences record the cumulative effect of the processes considering their arrangement in time.

Most of these principles are linked to the water budget and can be interpreted in terms of groundwater level evolution and, finally, to climate.

The presented sequences can be classified by the greater or lesser development of these features and, consequently, ranked according to the relative water budget during their formation. Thus, Sequences 1 and 3 are characterized by a lesser development of pedogenic features ("wet" sequences) whilst Sequences 2 and 4 show a greater development of such features ("arid" sequences).

Comparing Sequences 1 and 3, Sequence 1 contains pedogenic features both at the top and bottom (beginning of the sedimentation), and calcite presents as a chemical precipitate (marls upwards) but with low organic content. However, pedogenic features are only present at the top of Sequence 3, which contains more sandy layers and a greater organic content. Thus, it can be assumed that the climate during Sequence 3 was wetter than during Sequence 1.

As far as Sequences 2 and 4 are concerned, it can be observed that Sequence 4 presents chemical sedimentation (marls) and smaller pedogenic features than Sequence 2, but it does contain gypsum crystals. This is indicative of an overall less dry and warmer (higher evaporation rates) conditions for Sequence 4 than for Sequence 2.

There are few climate records from nearby areas with similar settings but, when they are compared (Figure 7), it is evident that there is a good correlation to other inland wetlands from the northern (Villalba de Adaja [56]) and southern Spanish Meseta (Tablas de Daimiel [57]), as well as to records from the surrounding mountains [58,59].

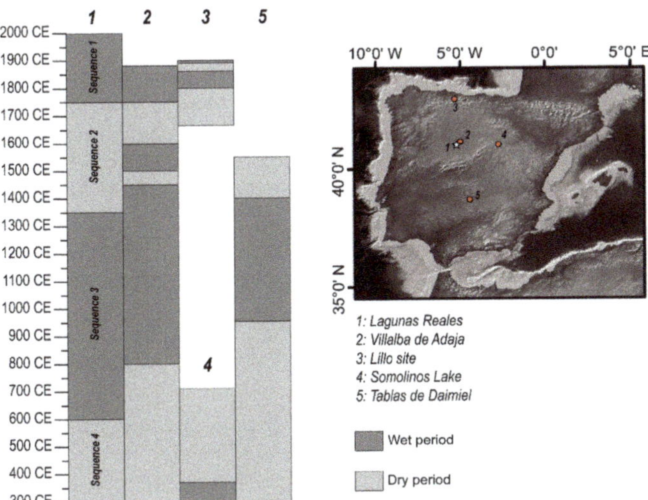

Figure 7. Hydroclimate periods for nearby records in the northern (1 to 4) and southern (5) Spanish Meseta.

For the most recent period (Sequence 1), the Villalba de Adaja pollen record accounts for warm and wet conditions, which agrees with the increase in rainfall documented for the last 150 years (Figure 2a) and the tree-ring-derived reconstruction of wet–dry conditions for the northern boundary of the northern Spanish Meseta [59].

For Sequence 2, the drier conditions evidenced by the sediments are correlative with the colder and drier signals captured by the tree rings [59] and pollen records from the northern [56] and southern [57] Spanish Meseta. Yet, historical sources document fishing in the Lagunas Reales as an economic activity, at least from 1610 to 1745 CE. This means that, despite the overall negative water balance for this period, during some years the water body was stable enough to allow fishing. An alternation of drought and flood periods has been documented [21,60–63] for this period, during the Little Ice Age.

However, the Medieval Climate Anomaly (Sequence 3) has been described as more hydrologically stable but spatially heterogeneous so wetter conditions were present in NW Spain and drier conditions in Mediterranean Spain [64]. Data from both Mesetas [56,57] agree with the wetter character of this period recorded in the Lagunas Reales, which should be considered within the "wetter" Spain for this period [64].

There are few records covering the period of Sequence 4 in inland Spain. This period, covering part of the so-called Roman Period plus the Early Medieval or Dark Ages, is characterized by arid conditions in the centre and southern margin of the northern [56,58] and southern Spanish Meseta [57].

The link between the oscillations of climate, groundwater and wetland sediments are usually found via the reconstruction of water levels. This reconstruction can be obtained by facies analysis (e.g., this paper and in [24,65,66]), which later will be equated to a water budget model (e.g., [25]). In other studies, the relationship between the water level and the water budget is obtained from paleo-salinity reconstructions (through paleo-ecological reconstructions or changes in the authigenic mineral or geochemical composition and/or isotope geochemistry [26,27,67–69]) that assume that these changes in salinity are due to changes in the evaporation/precipitation (E/P) ratio that affect the groundwater level.

In any case, these models assume that changes in salinity are related to changes in the E/P ratio acting on a non-changing groundwater mass and the processes are assumed to be in the time scale of the surface processes. These hypotheses have three caveats: (1) multilayer aquifers and mixing of different sources of water are a present day reality; (2) there is little knowledge about the relationship between the time scales of surface and under-surface processes, and there is a need to explore longer time scales than the decennial [70,71]; and (3) sediments contribute to groundwater chemistry [28,30,72,73], but groundwater also captures the signals of recharge chemistry due to climate and environmental features [74], whereas sediment records the features of the water [75].

The reconstruction of the paleo-salinity of wetland water has been a common technique but it depends on the assumption that increases in saline elements are related to evaporative processes on the surface and on the preservation of salts in the sediments. In addition, time scales of changes in the composition of groundwater are longer than those for changes in groundwater levels [75]. Therefore, the chemical signature is more stable than the water level one.

Three families of water have been identified in the surroundings of the Lagunas Reales, Cl^--Na^+, $HCO_3^--Cl^--Na^+-Ca^{2+}$ and $HCO_3^--Na^+-Ca^{2+}$ (Table 1), similar to those identified by [42] for the Lagunas de Villafáfila. These authors consider that the Cl^--Na^+ water are long and deep flow paths discharging into the ponds, the $HCO_3^--Ca^{2+}-Mg^{2+}$ ($HCO_3^--Na^+-Ca^{2+}$ in Lagunas Reales) to rainfall and surface water and the $HCO_3^--Cl^--Na^+$ ($HCO_3^--Cl^--Na^+-Ca^{2+}$ in Lagunas Reales) to a mixture of the previous ones.

As stated in Section 4.2, the geochemical composition of the sediments is related to different sources. Elements linked to the siliciclastic fraction carried by runoff into the pond (Si, Al, K, Ti and Fe), those linked to organic matter (U and P), Mn as related to redox conditions and a fourth set of elements (S, Cl, Br, Sr, Ca, Mg and As) that can be related to the long flow-path (long residence time) groundwater.

Thus, the sediments are recording surface (siliciclastic and organic matter supply by surface water), sub-surface (inputs from the underlying groundwater) and a mixture (evaporative processes of the water body) of processes that can help us to understand their interactions (Figure 8).

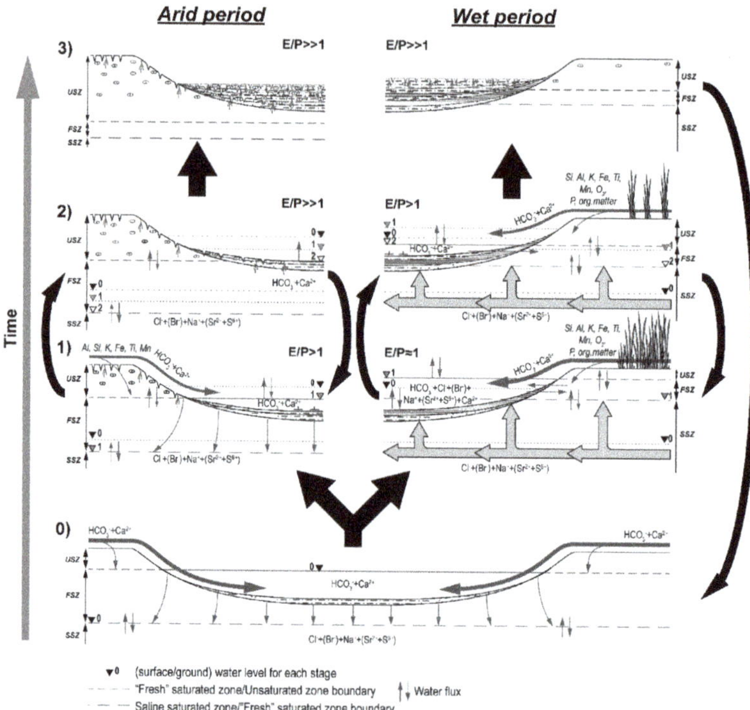

Figure 8. Conceptual model relating sedimentary facies, water chemistry and levels, as well as hydroclimate for the Lagunas Reales. USZ: unsaturated zone, FSZ: "fresh" water saturated zone, SSZ: "saline" water saturated zone. E/P: Evaporation/precipitation ratio. Sedimentary facies key as in Figure 3.

During arid periods (those where the overall water budget was negative in the long term), the long-term trend for the groundwater level is to fall as the regional water balance is negative. The wetland is filled by short-lived surface fresh water (runoff + rainfall, $HCO_3^--Na^+-Ca^{2+}$) and the wetland behaves as a recharge point for the aquifer through surface water (giving place to a temporary freshwater saturated zone, FSZ) whilst the deeper saline saturated zone (SSZ), related to the long path deeper groundwater, remains below the ground level. Reference [76] determined that summer infiltration in wetlands in Canada represent 70% of waters loss, recharging the groundwater under the wetlands. The unsaturated zone (USZ) is well developed and pedogenic processes, dominated by upward fluxes (E/P > 1), take place at the margins of the ponds. Vegetation is scarce due to the hydric stress, so organic matter inputs into the ponds are minor. During high water stages, the E/P > 1 ratio enhances the precipitation of calcite from the water body that mixes with the siliciclastic deposits carried by surface runoff. For the low water stages (E/P >> 1), the water table is below the surface and pedogenic processes (linked to pore water ascension by capillarity in the unsaturated zone) spread into the ponds giving place to the development of carbonate nodules and the colonization of the bottom of the pond by vegetation (root traces).

As these cycles are repeated throughout time, the progressive infill of the pond, in addition to the lowering of the water table, has led to the complete drying-out of the wetland for a prolonged period.

These processes explain the recorded sedimentary facies and the low content in siliciclastic and organic-derived elements (low surface runoff) and low groundwater-related elements (as the water table was low and the upper saturated zone filled by surface waters) for Sequences 2 and 4 (Figure 6).

During wet periods (those whose overall water budget was positive in the long term), the long-term trend of the groundwater level (SSZ + FSZ) is to rise as the regional water balance is positive. The wetland was a discharge area for the groundwater, leading to the development of a water body composed by a mixture of these waters and surface waters ($Cl^--Na^+ + HCO_3^--Na^+-Ca^{2+} = HCO_3^--Cl^--Na^+-Ca^{2+}$). There is an unsaturated zone (USZ) of variable thickness in time dominated by downward (E/P ≈ 1 or < 1) or upward (E/P > 1) fluxes, and a shallow freshwater saturated zone (FSZ) that promotes the development of marginal vegetation that, during runoff episodes, supplies organic matter, in addition to the siliciclastics, to the wetland. During high water stages (E/P ≈ 1 or < 1), the saline groundwater (SSZ) discharges into the ponds and the surface runoff supplies siliciclastics and organic matter to the wetland and freshwater to the ponds and groundwater (FSZ), resulting in a brackish water body. Clays and organic matter capture the groundwater-related elements (S, Sr, Cl and Br), and the availability of surface water (runoff + rainfall) prevents the formation of salts. For low water stages (E/P > 1), the saturated zone falls (FSZ + SSZ), but the water table (FSZ) remains above the ground surface; the change in chemistry of the water plus the increase in the E/P ratio promotes calcite precipitation. This situation is similar to the high-water stage of the arid periods.

Repetition of these cycles led to the described Sequence 3 (characterized by greater values of runoff and groundwater proxies and scarce development of pedogenic features in the sediment). The drying-out of this sequence is related to the progressive change to drier conditions in time (and the change to Sequence 2).

The arid period–wet period model shows the extremes of a continuum. Climate changes in a cyclic way so there are as many models as we might want. Sequence 1 represents an intermediate model as it shows features common to the arid period model (low values of surface runoff and groundwater proxies) but some others point to wetter conditions than those represented for that model (increasing values for runoff and groundwater proxies, and noticeable amounts of calcite precipitation that can be linked to both models).

This model introduces a change as compared to classical approaches [25,26,68] as they considered that increases in salinity recorded in lakes and wetlands were only related to greater evaporation rates, drier conditions. The Lagunas Reales example shows that an increase in saline elements can be related to wetter conditions as they are linked to changes in the mixture of surface and deep waters due to changes in the groundwater levels.

It provides a new approach to study the long-term relationships beyond the decennial and centennial time scales, although there still are some points that need further study. Some authors [70,71] point out that the signal lagging and damping between the climate and groundwater response is poorly known, and this is a key point to apply this knowledge in water management and climate studies. This could be achieved by enhancing the time resolution of the sediment studies; by improving the age models via more dating; a better identification of the climate signal, which can be obtained by increasing the geochemical (isotopes) and fossil proxies; and the integration of these studies into water models.

6. Conclusions

Wetlands are very valuable environments, and very sensitive to water balances and, therefore, to climate change and human action. The knowledge of their behaviour and the relationship between surface water and groundwater and climate can be a key not only for their preservation but also to understand the hydrological cycle. Several authors have stressed the relevance of their sediments to understand past hydrological and climate change (i.e., groundwater-discharge deposits [77]).

Groundwater takes its chemical fingerprint from the rocks that it crosses, but sediments deposited in this process (discharge areas) record these compositions. The use of geochemical proxies related to surface water (rainfall + runoff: siliciclastics + organic matter) and groundwater (enriched in saline elements resulting from long-path flows) has allowed us to study the relationship between both sources of water through time.

The resulting models for these small saline lakes show that wet periods are characterized by high groundwater levels that allow the saline deeper water to arise on the surface and leave its fingerprint in the geochemistry of the sediments, whilst dry periods are characterized by lower groundwater levels and a greater influence of surface water (less saline) that are also recorded in the sediments.

Thus, in this example, the saline fingerprint is not related to more evaporation but rather to higher groundwater levels during wet periods. This introduces a new idea that must be considered when reconstructing past water levels and climate from groundwater-related deposits.

Author Contributions: Conceptualization, R.M. and J.I.S.; data curation, R.M., J.I.S., I.L.-C., L.G.d.F. and Á.D.l.H.-P.; investigation, R.M., J.I.S., I.L.-C. and L.G.d.F.; methodology, R.M. and J.I.S.; project administration, Á.d.l.H.-P.; visualization, J.I.S.; writing—original draft, R.M., J.I.S. and I.L.-C.; writing—review and editing, L.G.d.F. and Á.d.l.H.-P. All authors have read and agreed to the published version of the manuscript.

Funding: This research has received funding from the European Union H2020 Programme under Grant agreement No. 730497 for the research project NAIAD (NAture Insurance value: Assessment and Demonstration). IGME equipment is co-financed by the European Regional Development Fund via the Spanish National Plan for Scientific and Technical Research and Innovation 2013–2016 (IGME13-4E-2576).

Acknowledgments: The authors are very grateful to Angela Tate for the English proofreading. We are also very thankful to the anonymous reviewers and the academic editor whose comments and suggestions have helped us to improve the manuscript.

References

1. Ramsar Convention Secretariat. *An Introduction to the Convention on Wetlands (Previously the Ramsar Convention Manual)*, 5th ed.; Ramsar Convention Secretariat: Gland, Switzerland, 2016.

2. Mitra, S.; Wassmann, R.; Vlek, P.L.G. *Global Inventory of Wetlands and Their Role in the Carbon Cycle, No. 64*; ZEF Discussion Papers on Development Policy, Center for Development Research (ZEF): Bonn, Germany, 2003; pp. 1–44.

3. Russi, D.; Brink, P.; Farmer, A.; Badura, T.; Coates, D.; Förster, J.; Kumar, R.; Davidson, N. *The Economics of Ecosystems and Biodiversity for Water and Wetlands*; IEEP: London, UK, 2013; pp. 1–78.

4. Stacke, T.; Hagemann, S. Development and validation of a global dynamical wetlands extent scheme. *Hydrol. Earth Syst. Sci.* **2012**, *16*, 2915–2933. [CrossRef]

5. Davidson, N.C. How much wetland has the world lost? Long-term and recent trends in global wetland area. *Mar. Freshw. Res.* **2014**, *65*, 934–941. [CrossRef]

6. Comín, F.A.; Alonso, M. Spanish salt lakes: Their chemistry and biota. *Hydrobiologia* **1988**, *158*, 237–245. [CrossRef]

7. Casado, S.; Montes, C. *Guía de los Lagos y Humedales de España*; J.M. Reyero: Madrid, Spain, 1995; pp. 1–255.

8. Santisteban, J.I.; Mediavilla, R.; Domínguez-Castro, F.; Gil-García, M.J.; Ruiz-Zapata, M.B.; Dabrio, C.J. From a climate to man-controlled C budget in a temperate wetland (Las Tablas de Daimiel National Park, Central Spain). In *Earth: Our Changing Planet, Proceedings of IUGG XXIV General Assembly, Perugia, Italy, 2–13 July 2007*; Ubertini, L., Manciola, P., Casadei, S., Grimaldi, S., Eds.; International Union of Geodesy and Geophysics: Perugia, Italy, 2007; p. 1257.

9. Höbig, N.; Mediavilla, R.; Gibert, L.; Santisteban, J.I.; Cendon, D.I.; Ibáñez, J.; Reicherter, K. Palaeohydrological evolution and implications for paleoclimate since the Late Glacial at Laguna de Fuente de Piedra, southern Spain. *Quat. Int.* **2016**, *407*, 29–46. [CrossRef]

10. Reed, J.M.; Stevenson, A.C.; Juggins, S. A multi-proxy record of Holocene climatic change in southwestern Spain: The Laguna Medina, Cádiz. *Holocene* **2001**, *11*, 707–719. [CrossRef]

11. Martín-Puertas, C.; Valero-Garcés, B.L.; Mata, M.P.; González-Samperiz, P.; Bao, R.; Moreno, A.; Stefanova, V. Arid and humid phases in southern Spainduring the last 4000 years: The Zoñar Lake record, Córdoba. *Holocene* **2008**, *18*, 907–921. [CrossRef]

12. Martín-Puertas, C.; Jiménez-Espejo, F.; Martínez-Ruiz, F.; Nieto-Moreno, V.; Rodrigo, M.; Mata, M.P.; Valero-Garcés, B.L. Late Holocene climate variability in the southwestern Mediterranean region: An integrated marine and terrestrial geochemical approach. *Clim. Past* **2010**, *6*, 807–816. [CrossRef]

13. Valero-Garcés, B.L.; Delgado-Huertas, A.; Navas, A.; Machín, J.; González-Sampériz, P.; Kelts, K. Quaternary palaeohydrological evolution of the playa lake: Salada Mediana, central Ebro Basin, Spain. *Sedimentology* **2000**, *47*, 1135–1156. [CrossRef]

14. Pérez, A.; Luzón, A.; Roc, A.; Soria, A.; Mayayo, M.; Sánchez, J.A. Sedimentary facies distribution and genesis of a recent carbonate-rich saline lake: Gallocanta Lake. Iberian Chain, NE Spain. *Sediment. Geol.* **2002**, *148*, 185–202. [CrossRef]

15. Luzón, A.; Pérez, A.; Mayayo, M.J.; Soria, A.R.; Sánchez, J.A.; Roc, A.C. Palaeogeographical changes since 11,000 yr BP in the Gallocanta lacustrine basin. Iberian Range. NE Spain. *Holocene* **2007**, *17*, 649–663. [CrossRef]

16. Santisteban, J.I.; García del Cura, M.A.; Mediavilla, R.; Dabrio, C.J. Estudio preliminar de los sedimentos recientes de las Lagunas de Villafáfila (Zamora). *Geogaceta* **2003**, *33*, 51–54.

17. López-Sáez, J.A.; Abel-Schaad, D.; Iriarte, E.; Alba-Sánchez, F.; Pérez-Díaz, S.; Guerra-Doce, E.; Delibes de Castro, G.; Abarquero Moras, F.J. Una perspectiva paleoambiental de la explotación de la sal en las Lagunas de Villafáfila (Tierra de Campos, Zamora). *Cuatern. Geomorfol.* **2017**, *31*, 73–104. [CrossRef]

18. Sánchez-López, G.; Hernández, A.; Pla-Rabes, S.; Trigo, R.M.; Toro, M.; Granados, I.; Sáez, A.; Masqué, P.; Pueyo, J.; Rubio-Inglés, M.J.; et al. Climate reconstruction for the last two millennia in central Iberia: The role of East Atlantic (EA), North Atlantic Oscillation (NAO) and their interplay over the Iberian Peninsula. *Quat. Sci. Rev.* **2016**, *149*, 135–150. [CrossRef]

19. Giralt, S.; Moreno, A.; Cacho, I.; Valero-Garcés, B.L. A comprehensive overview of the last 2000 years Iberian Peninsula climate history. *CLIVAR Exchanges* **2017**, *73*, 5–10.

20. Valero-Garcés, B.L.; González-Sampériz, P.; Navas, A.; Machín, J.; Mata, P.; Delgado-Huertas, A.; Bao, R.; Moreno, A.; Carrión, J.S.; Schwalb, A.; et al. Human impact since medieval times and recent ecological restoration in a Mediterranean lake: The Laguna Zoñar, southern Spain. *J. Paleolimnol.* **2006**, *35*, 441–465. [CrossRef]

21. Santisteban, J.I.; Mediavilla, R.; Gil-García, M.J.; Domínguez-Castro, F.; Ruiz-Zapata, M.B. La historia a través de los sedimentos: Cambios climáticos y de usos del suelo en el registro reciente de un humedal mediterráneo (Las Tablas de Daimiel, Ciudad Real). *Boletín Geológico Min.* **2009**, *120*, 497–508.

22. Dubois, N.; Saulnier-Talbot, E.; Mills, K.; Gell, P.; Battarbee, R.; Bennion, H.; Chawchai, S.; Dong, X.; Francus, P.; Flower, R.; et al. First human impacts and responses of aquatic systems: A review of palaeolimnological records from around the world. *Anthr. Rev.* **2018**, *5*, 28–67. [CrossRef]

23. Fuentealba, M.; Latorre, C.; Frugone-Álvarez, M.; Sarricolea, P.; Giralt, S.; Contreras-López, M.; Prego, R.; Bernárdez, P.; Valero-Garcés, B. A combined approach to establishing the timing and magnitude of anthropogenic nutrient alteration in a Mediterranean coastal lake watershed system. *Sci. Rep.* **2020**, *10*, 5864. [CrossRef]

24. Digerfeldt, G.; Almendinger, J.E.; Björk, S. Reconstruction of past lake levels and their relation to groundwater hydrology in the Parkers Prairie sandplain, west-central Minnesota. *Palaeogeogr. Palaeoclimatol. Palaeoecol.* **1992**, *94*, 99–118. [CrossRef]

25. Almendinger, J.E. A groundwater model to explain past lake levels at Parkers Prairie, Minnesota, USA. *Holocene* **1993**, *3*, 105–115. [CrossRef]

26. Donovan, J.J.; Smith, A.J.; Panek, V.A.; Engstrom, D.R.; Ito, E. Climate-driven hydrologic transients in lake sediment records: Calibration of groundwater conditions using 20th Century drought. *Quat. Sci. Rev.* **2002**, *21*, 605–624. [CrossRef]

27. Fritz, S.C. Deciphering climatic history from lake sediments. *J. Paleolimnol.* **2008**, *39*, 5–16. [CrossRef]

28. Nissenbaum, A.; Magaritz, M. Bromine-Rich Groundwater in the Hula Valley, Israel. *Naturwissenschaften* **1991**, *78*, 217–218. [CrossRef]

29. Gómez, J.J.; Lillo, J.; Sahún, B. Naturally occurring arsenic in groundwater and identification of the geochemical sources in the Duero Cenozoic Basin, Spain. *Environ. Geol.* **2006**, *50*, 1151–1170. [CrossRef]

30. Gwynne, R.; Frape, S.; Shouakar-Stash, O.; Love, A. ^{81}Br, ^{37}Cl, and ^{87}Sr studies to assess groundwater flow and solute sources in the southwestern Great Artesian Basin, Australia. *Procedia Earth Planet. Sci.* **2013**, *7*, 330–333. [CrossRef]

31. Pineda Velasco, A.; Salazar Rincón, A.; Herrero Hernández, A.; Camero Benito, Y. *Mapa Geológico de España, E. 1:50.000, 2ª Serie (MAGNA). Hoja 427 (Medina del Campo)*; Instituto Tecnológico Geominero de España: Madrid, Spain, 2007; pp. 1–75.

32. Fernández Pereira, J.; Rodríguez Arroyo, J.; del Barrio, V.; Ramos, M.A.; Castrillón, M.; Vaquerizo, E.; Trujillo, H.; Hernández, V.; Gómez, S.; Seisdedos, P.; et al. *Plan Hidrológico de la parte española de la demarcación hidrográfica del Duero (2015–2021)*; Memoria. Ministerio de Agricultura, Alimentación y Medio Ambiente, Confederación Hidrográfica del Duero: Valladolid, Spain, 2015; pp. 1–486.

33. Nafría García, D.A.; Garrido del Pozo, N.; Álvarez Arias, M.V.; Cubero Jiménez, D.; Fernández Sánchez, M.; VIllarino Barrera, I.; Gutiérrez García, A.; Abia Llera, I. *Atlas Agroclimático de Castilla y León*; Agencia Estatal de Meteorología: Madrid, Spain, 2013; pp. 1–135.

34. Mediavilla, R.; Santisteban, J.I.; Borruel-Abadía, V.; López-Cilla, I. Chapter 6. Past and recent sedimentation at the Lagunas Reales and their long-term evolution. In *Medina del Campo Case Study*; Confederación Hidrográfica del Duero: Valladolid, Spain, under review.

35. IGME-DGA. *Apoyo a la Caracterización Adicional de las Masas de Agua Subterránea en Riesgo de no Cumplir los Objetivos Medioambientales en 2015. Demarcación Hidrográfica del Duero. Masa de Agua Subterránea 47 Medina del Campo*; Ministerio de Ciencia e Innovación/Ministerio de Medio Ambiente y Medio Rural y Marino: Madrid, Spain, 2010; pp. 1–39.

36. Alcalá, F.J.; Custodio, E. Using the Cl/Br ratio as a tracer to identify the origin of salinity in aquifers in Spain and Portugal. *J. Hydrol.* **2008**, *359*, 189–207. [CrossRef]

37. Gonzalez Bernáldez, F.; Herrera, P.; Sastre, A.; Rey, J.M.; Vicente, R. Comparación preliminar de los ecosistemas de descarga de aguas subterráneas en las cuencas del Duero y del Tajo. *Hidrogeol. Recur. Hidráulicos* **1987**, *11*, 19–39.

38. González Bernáldez, F. Ecological aspects of wetland/groundwater relationships in Spain. *Limnetica* **1992**, *8*, 11–26.

39. Rey Benayas, J.M.; Pérez, C.; González Bernáldez, F.; Zabaleta, M. Tipología y cartografía de los humedales de las cuencas del Duero y Tajo. *Mediterr. Sec. Biol.* **1990**, *12*, 5–26.

40. Rey Benayas, J.M.; González Bernáldez, F.; Levassor, C.; Peco, B. Vegetation of groundwater discharge sites in the Douro basin, central Spain. *J. Veg. Sci.* **1990**, *1*, 461–466. [CrossRef]

41. Fernández Pérez, L.; Cabrera Lagunilla, M.P. Estudio hidrogeológico de las Lagunas de Villafáfila. In *Proceedings of the III Reunión Nacional sobre Geología Ambiental y Ordenación del Territorio, Valencia, Spain*; Comunicaciones, I; Serv. Publicaciones Universidad de Valencia: Valencia, Spain, 1987; pp. 441–459.

42. Armenteros, I.; Huerta, P.; Cidón-Trigo, A.; Rueda-Gualdrón, M.C.; Recio, C.; Martínez-Graña, A. Hidrogeología del entorno de las Lagunas de Villafáfila (Zamora). *Geogaceta* **2019**, *66*, 51–54.

43. Cabestrero, O.; Sanz-Montero, M.E. Brine evolution in two inland evaporative environments: Influence of microbial mats in mineral precipitation. *J. Paleolimnol.* **2016**, *59*, 139–157. [CrossRef]

44. Bronk Ramsey, C. Bayesian analysis of radiocarbon dates. *Radiocarbon* **2009**, *51*, 337–360. [CrossRef]

45. Reimer, P.J.; Bard, E.; Bayliss, A.; Beck, J.W.; Blackwell, P.G.; Bronk Ramsey, C.; Grootes, P.M.; Guilderson, T.P.; Haflidason, H.; Hajdas, I.; et al. IntCal13 and Marine13 Radiocarbon Age Calibration Curves 0–50,000 Years cal BP. *Radiocarbon* **2013**, *55*, 1869–1887. [CrossRef]

46. Calvert, S.E.; Pedersen, T.F. Chapter Fourteen Elemental Proxies for Palaeoclimatic and Palaeoceanographic Variability in Marine Sediments: Interpretation and Application. In *Developments in Marine Geology*; Hillaire-Marcel, C., De Vernal, A., Eds.; Elsevier: Amsterdam, The Netherlands, 2007; Volume 1, pp. 567–644.

47. Santisteban, J.I.; Mediavilla, R.; Galán de Frutos, L.; López Cilla, I. Holocene floods in a complex fluvial wetland in central Spain: Environmental variability, climate and time. *Glob. Planet. Chang.* **2019**, *181*, 102986. [CrossRef]

48. Engstrom, D.R.; Hansen, B.C.S. Postglacial vegetational change and soil development in southeastern Labrador as inferred from pollen and chemical stratigraphy. *Can. J. Bot.* **1985**, *63*, 543–561. [CrossRef]

49. Naeher, S.; Gilli, A.; North, R.P.; Hamann, Y.; Schubert, C.J. Tracing bottom water oxygenation with sedimentary Mn/Fe ratios in Lake Zurich, Switzerland. *Chem. Geol.* **2013**, *352*, 125–133. [CrossRef]

50. R Core Team. R. *A Language and Environment for Statistical Computing*; R Foundation for Statistical Computing: Vienna, Austria, 2020.

51. Freytet, P.; Plaziat, J.C. *Continental Carbonate Sedimentation and Pedogenesis—Late Cretaceous and Early Tertiary of Southern France*; Schweizerbart Science Publishers: Stuttgart, Germany, 1982; pp. 1–213.

52. Wright, V.P. Terrestrial stromatolites and laminar calcretes: A review. *Sediment. Geol.* **1989**, *65*, 1–13. [CrossRef]

53. Mack, G.H.; James, W.C. Paleoclimate and the Global Distribution of Paleosols. *J. Geol.* **1994**, *102*, 360–366. [CrossRef]

54. Kraus, M.J.; Aslan, A. Eocene hydromorphic Paleosols; significance for interpreting ancient floodplain processes. *J. Sediment. Res.* **1993**, *63*, 453–463.

55. Pipujol, M.D.; Buurman, P. Dynamics of iron and calcium carbonate redistribution and palaeohydrology in middle Eocene alluvial paleosols of the southeast Ebro Basin margin (Catalonia, northeast Spain). *Palaeogeogr. Palaeoclimatol. Palaeoecol.* **1997**, *134*, 87–107. [CrossRef]

56. López Merino, L.; López Sáez, J.A.; Alba Sánchez, F.; Pérez Díaz, S.; Abel Schaad, D.; Guerra Doce, E. Estudio polínico de una laguna endorreica en Almenara de Adaja (Valladolid, Meseta Norte): Cambios ambientales y actividad humana durante los últimos 2800 años. *Rev. Española Micropaleontol.* **2009**, *41*, 333–347.

57. Gil-García, M.J.; Ruiz Zapata, M.B.; Santisteban, J.I.; Mediavilla, R.; López-Pamo, E.; Dabrio, C.J. Late Holocene environments in Las Tablas de Daimiel (south central Iberian Peninsula, Spain). *Veg. Hist. Archaeobot.* **2007**, *16*, 241–250. [CrossRef]

58. Currás, A.; Zamora, L.; Reed, J.M.; García-Soto, E.; Ferrero, S.; Armengol, X.; Mezquita-Joanes, F.; Marqués, M.A.; Riera, S.; Julià, R. Climate change and human impact in central Spain during Roman times: High-resolution multi-proxy analysis of a tufa lake record (Somolinos, 1280 m asl). *Catena* **2012**, *89*, 31–53. [CrossRef]

59. Andreu-Hayles, L.; Ummenhofer, C.C.; Barriendos, M.; Schleser, G.H.; Helle, G.; Leuenberger, M.; Gutiérrez, E.; Cook, E.R. 400 Years of summer hydroclimate from stable isotopes in Iberian trees. *Clim. Dyn.* **2017**, *49*, 143–161. [CrossRef]

60. Grove, A.T. The "Little Ice Age" and its geomorphological consequences in Mediterranean Europe. *Clim. Chang.* **2001**, *48*, 121–136. [CrossRef]

61. Domínguez-Castro, F.; Santisteban, J.I.; Barriendos, M.; Mediavilla, R. Reconstruction of drought episodes for central Spain from rogation ceremonies recorded at the Toledo Cathedral from 1506 to 1900: A methodological approach. *Glob. Planet. Chang.* **2008**, *63*, 230–242. [CrossRef]

62. Loisel, J.; MacDonald, G.M.; Thomson, M.J. Little Ice Age climatic erraticism as an analogue for future enhanced hydroclimatic variability across the American Southwest. *PLoS ONE* **2017**, *12*, e0186282. [CrossRef]

63. Oliva, M.; Ruiz-Fernández, J.; Barriendos, M.; Benito, G.; Cuadrat, J.M.; Domínguez-Castro, F.; Garcia-Ruiz, J.M.; Giralt, S.; Gómez-Ortiz, A.; Hernández, A.; et al. The Little Ice Age in Iberian mountains. *Earth-Sci. Rev.* **2018**, *177*, 175–208. [CrossRef]

64. Moreno, A.; Pérez, A.; Frigola, J.; Nieto-Moreno, V.; Rodrigo-Gámiz, M.; Martrat, B.; González-Sampériz, P.; Morellón, M.; Martín-Puertas, C.; Corella, J.P.; et al. The medieval climate Anomaly in the Iberian Peninsula reconstructed from marine and lake records. *Quat. Sci. Rev.* **2012**, *43*, 16–32. [CrossRef]

65. Digerfeldt, G. Reconstruction and regional correlation of Holocene lake-level fluctuations in Lake Bysjön, South Sweden. *Boreas* **1988**, *17*, 165–182. [CrossRef]

66. Harrison, S.P.; Digerfeldt, G. European lakes as palaeohydrological and palaeoclimatic indicators. *Quat. Sci. Rev.* **1993**, *12*, 233–248. [CrossRef]

67. Telford, R.J.; Lamb, H.F.; Umer Mohammed, M. Diatom-derived palaeoconductivity estimates for Lake Awassa, Ethiopia: Evidence for pulsed inflows of saline groundwater? *J. Paleolimnol.* **1999**, *21*, 409–421. [CrossRef]

68. Fritz, S.C.; Ito, E.; Yu, Z.; Laird, K.R.; Engstrom, D.R. Hydrologic Variation in the Northern Great Plains During the Last Two Millennia. *Quat. Res.* **2000**, *53*, 175–184. [CrossRef]

69. Finney, B.P.; Bigelow, N.H.; Barber, V.A.; Edwards, M.E. Holocene climate change and carbon cycling in a groundwater-fed, boreal forest lake: Dune Lake, Alaska. *J. Paleolimnol.* **2012**, *48*, 43–54. [CrossRef]

70. Li, B.; Rodell, M. Evaluation of a model-based groundwater drought indicator in the conterminous U.S. *J. Hydrol.* **2015**, *526*, 78–88. [CrossRef]

71. Rust, W.; Holman, I.; Corstanje, R.; Bloomfield, J.; Cuthbert, K. A conceptual model for climate teleconnection signal control on groundwater variability in Europe. *Earth-Sci. Rev.* **2018**, *177*, 164–174. [CrossRef]

72. Sø, H.U.; Postma, D.; Lan, V.M.; Trang, P.T.K.; Kazmierczak, J.; Nga, D.V.; Pi, K.; Koch, C.B.; Viet, P.H.; Jakobsen, R. Arsenic in Holocene aquifers of the Red River floodplain, Vietnam: Effects of sediment-water interactions, sediment burial age and groundwater residence time. *Geochim. Cosmochim. Acta* **2018**, *225*, 192–209. [CrossRef]

73. Wang, Y.; Pi, K.; Fendorf, S.; Deng, Y.; Xie, X. Sedimentogenesis and hydrobiogeochemistry of high arsenic Late Pleistocene-Holocene aquifer systems. *Earth-Sci. Rev.* **2019**, *189*, 79–98. [CrossRef]

74. Cook, P.G.; Böhlke, J.K. Determining Timescales for Groundwater Flow and Solute Transport. In *Environmental Tracers in Subsurface Hydrology*; Cook, P.G., Herczeg, A.L., Eds.; Springer: Boston, MA, USA, 2000; pp. 1–30.

75. Custodio, E. Las aguas subterráneas como elemento básico de la existencia de numerosos humedales. *Ing. Agua* **2010**, *17*, 119–135. [CrossRef]

76. Parsons, D.F.; Hayashi, M.; van der Kamp, G. Infiltration and solute transport under a seasonal wetland: Bromide tracer experiments in Saskatoon, Canada. *Hydrol. Process.* **2002**, *18*, 2011–2027. [CrossRef]

77. Pigati, J.S.; Rech, J.A.; Quade, J.; Bright, J. Desert wetlands in the geologic record. *Earth-Sci. Rev.* **2014**, *132*, 67–81. [CrossRef]

Article

Using the Turnover Time Index to Identify Potential Strategic Groundwater Resources to Manage Droughts within Continental Spain

David Pulido-Velazquez [1,*], Javier Romero [2], Antonio-Juan Collados-Lara [1], Francisco J. Alcalá [3,4], Francisca Fernández-Chacón [5] and Leticia Baena-Ruiz [1]

[1]   Instituto Geológico y Minero de España, Urb. Alcázar del Genil, 4. Edificio Zulema, Bajo, 18006 Granada, Spain; aj.collados@igme.es (A.-J.C.-L.); l.baena@igme.es (L.B.-R.)
[2]   Campus de los Jerónimos s/n, Universidad Católica San Antonio de Murcia, Guadalupe, 30107 Murcia, Spain; romerog.javier@gmail.com
[3]   Instituto Geológico y Minero de España, Ríos Rosas, 23, 28003 Madrid, Spain; fj.alcala@igme.es
[4]   Instituto de Ciencias Químicas Aplicadas, Facultad de Ingeniería, Universidad Autónoma de Chile, Santiago 7500138, Chile
[5]   IES Ribera de Fardes, Cerro de los Almendrillos, Purullena, 18519 Granada, Spain; paquifchacon@gmail.com
*   Correspondence: d.pulido@igme.es; Tel.: +34-(95)-818-3143

Received: 1 October 2020; Accepted: 18 November 2020; Published: 22 November 2020

Abstract: The management of droughts is a challenging issue, especially in water scarcity areas, where this problem will be exacerbated in the future. The aim of this paper is to identify potential groundwater (GW) bodies with reduced vulnerability to pumping, which can be used as buffer values to define sustainable conjunctive use management during droughts. Assuming that the long term natural mean reserves are maintained, a preliminary assessment of GW vulnerability can be obtained by using the natural turnover time (T) index, defined in each GW body as the storage capacity (S) divided by the recharge (R). Aquifers where R is close to S are extremely vulnerable to exploitation. This approach will be applied in the 146 Spanish GW bodies at risk of not achieving the Water Framework Directive (WFD objectives, to maintain a good quantitative status. The analyses will be focused on the impacts of the climate drivers on the mean T value for Historical and potential future scenarios, assuming that the Land Use and Land Cover (LULC) changes and the management strategies will allow maintenance of the long term mean natural GW body reserves. Around 26.9% of these GW bodies show low vulnerability to pumping, when viewing historical T values over 100 years, this percentage growing to 33.1% in near future horizon values (until 2045). The results show a significant heterogeneity. The range of variability for the historical T values is around 3700 years, which also increases in the near future to 4200 years. These T indices will change in future horizons, and, therefore, the potential of GW resources to undergo sustainable strategies to adapt to climate change will also change accordingly, making it necessary to apply adaptive management strategies.

Keywords: drought; vulnerability to pumping; residence time; conjunctive use; sustainable management; climate change; adaptation strategies; Spanish GW bodies in quantitative risk

---

1. Introduction

The management of droughts is a challenging issue, especially in water scarcity areas with water deficits in terms of long-term average conditions [1]. These deficits can be observed in precipitation, soil moisture, river discharge or supply in relation to water demands, defining respectively different drought types: meteorological, agricultural, hydrological, or even operational droughts depending on the variables used to assess them [2]. For example, operational drought or scarcity [3] is related to the deficit in water demand satisfaction in a system. In most of these areas, the frequency and intensity of the drought events will be exacerbated in the future, due to climate change [4]. In this framework, groundwater (GW) may play a significant role for sustainable management of water scarcity, due to its role as buffer value, providing additional resources that can be temporarily employed to cover necessities during critical droughts [5]. GW resources are also crucial for appropriate analyses of scarcity, and due to aquifer status have an important influence on water availability to fulfil demands. GW overexploitation is an issue with even higher impacts in lowering GW levels than climate change in many regions, especially in the Mediterranean area [6]. These impacts have been explicitly analysed and discussed in research works on coastal [7] and non-coastal Mediterranean aquifers [8]. Drought also exacerbates aquifer overexploitation, a significant issue in the Mediterranean area [9].

Increased water availability has resulted in an even larger increase in water supply demands, as demonstrated in Tunisia by Besbes et al. [10]. Groundwater that was frequently exploited as a consequence of the silent revolution in agricultural development gave rise to a demand for irrigation which accounts for 60–80% of the total water demand in the Mediterranean area today. As a result of this highly exacerbated groundwater use, water tables in some regions have fallen by as much as several hundred meters, such as in southeast Spain [11]. Among other significant immediate effects, seawater intrusion into coastal aquifers has also been recorded [12]. Notwithstanding the numerous interpretations of overexploitation as a general term [13], it is hereinafter used to describe the exhaustive use of groundwater resources that pose a subsequent risk to the preservation of groundwater quality and quantity in the long-term, and/or to additional services provided by the aquifers.

On the other hand, the legal EU water management context defined by the Water Framework Directive [14] aims to achieve a sustainable management of the resources, maintaining a good status of surface and GW bodies. The state members, for the different planning horizons, have identified Water Bodies at risk of not achieving the Water Framework Directive (WFD) (2000) [14] objectives and have proposed programs oriented to fulfill these targets. Therefore, in the decision-making process when managing water resource systems, special attention should be paid to the management of these GW bodies.

The concept of vulnerability is closely related to GW body status and risks. It has been extensively studied from the perspective of vulnerability to surface pollution. Different approaches and techniques [15] can be applied to assess both intrinsic and specific vulnerability. Intrinsic vulnerability focuses on analyses of the ease with which any surface pollution can reach and extend within the aquifer saturated zone [16]. Specific vulnerability refers to a particular contaminant or groups of contaminants, taking into account their properties and the potential processes and interactions that may influence them [17]. From this qualitative point of view, the vulnerability of a GW-resource to pollution depends on intrinsic susceptibility, which depends on the aquifer properties, the associated sources of water, the distribution and types of contamination sources (natural and/or anthropogenic), and the transport of the contaminants [18]. The most commonly employed methods to assess vulnerability are the index-based approaches [8] that include methods such as DRASTIC [19], Aquifer vulnerability index (AVI) [20] and SINTACS [21] or adaptation of them. In the Mediterranean region, SINTACS and SINTACS-derived methods are the most commonly applied, both for intrinsic and specific vulnerability assessment [22]. This GW vulnerability concept has also been linked to variable GW ages, travel and residence times.

GW age can be estimated by using environmental and isotope tracers [23,24] and model simulations [25,26]. It has been related to vulnerability of production wells to contamination in some cases [27]. The mean age of the water leaving the system or the mean residence time is also known as the mean GW age or renewal time, and can be approached, under natural conditions, by the mean natural GW turnover time (T) index, which is obtained by dividing the total storage capacity (S) by the net GW recharge (R) [23]. Therefore, the value of this index in the future could be affected by climate change scenarios and their impacts on rainfall aquifer recharge [28–30].

In this paper we propose a novel method to perform a preliminary analysis of GW vulnerability to intensive pumping during drought periods, assuming that the long term natural mean reserves are maintained by the actual recharge of the main inflow of groundwater resources. This will be analyzed under historical and potential future climate scenarios. The pumping vulnerability concept is introduced and assessed by applying a T index approach. This allows us to identify potential strategic GW bodies for sustainable conjunctive use management of critical droughts in water scarcity areas in continental Spain. We also studied the significance and variability of the R and S variables employed to obtain the T index depending on the aquifer lithology. Finally, we analyzed whether some potential explanatory variables could be employed to describe the T distribution. Due to T dependencies on R, which can be estimated from the effective precipitation (precipitation minus the actual evapotranspiration), by applying an effective recharge coefficient (C), the explanatory analyses were also extended to the recharge coefficient.

2. Materials and Methods

2.1. Methodology

This novel method intends to perform a preliminary analysis of GW vulnerability to intensive pumping during drought periods through the renewal time of resources (GW age), approached by the T index as the S/R ratio. Assuming that the long term natural mean reserves are kept invariant and the actual recharge is the main inflow of groundwater resources, the GW bodies with high renewal time will be less vulnerable to pumping than those with low values, even in periods in which pumping is smaller than mean R. This can be especially relevant in Basins or Water Resource systems with scarce reserves where long and intensive droughts appear and will be exacerbated in the future due to climate change. The methodology is summarized in Figure 1.

Making a parallel between unconfined aquifers and reservoirs, the GW discharge (Q) will start when the potential aquifer storage reaches the threshold level of the surface connection (see Figure 1). Assuming that there is no pumping, a preliminary assessment of the natural mean age of the groundwater leaving the GB body through the connection with the surface system (springs and or stream-aquifer interaction boundary conditions) can be obtained through the natural mean T index, defined as:

$$T = S/R \tag{1}$$

where T, S, and R are defined in the caption to Figure 1.

In each GW body, S can be obtained by combining information about the geometry and the storage coefficients, which can be derived from different sources (e.g., field works, models and/or research papers and official reports, as well as the River Basin Plans published by the different River Basin authorities). The historical R can be estimated through field work or previously calibrated models [29,31,32]. The historical mean T value can be estimated by combining the mean historical R values with S in accordance with Equation (1).

The impacts of future potential climatic scenarios on GW bodies R, and, therefore, on their T index, requires climatic scenarios to be downscaled and propagated with a previously calibrated recharge model. In order to generate future local scenarios from the simulations of the last future emission scenarios defined by the Intergovernmental Panel on Climate Change (IPCC) (AR5) with climatic models, we need to correct these with local historical climatic data for the case study [33]. Two different statistical approaches may be applied to perform the correction or downscaling, a delta change approach and/or a bias correction approach [34]. Local scenarios can be generated for a future horizon or a specific global warming level (e.g., 1 or 2 °C above the reference period) [35]. We propose to define an equi-feasible ensemble of multiple local projections, which are supposed to produce more robust and representative scenarios than those based on single projections [36]. The impacts of the generated potential local scenarios on the mean T will require future mean R to be estimated, which will be assessed by propagating/simulating the generated climatic scenarios with previously calibrated recharge models.

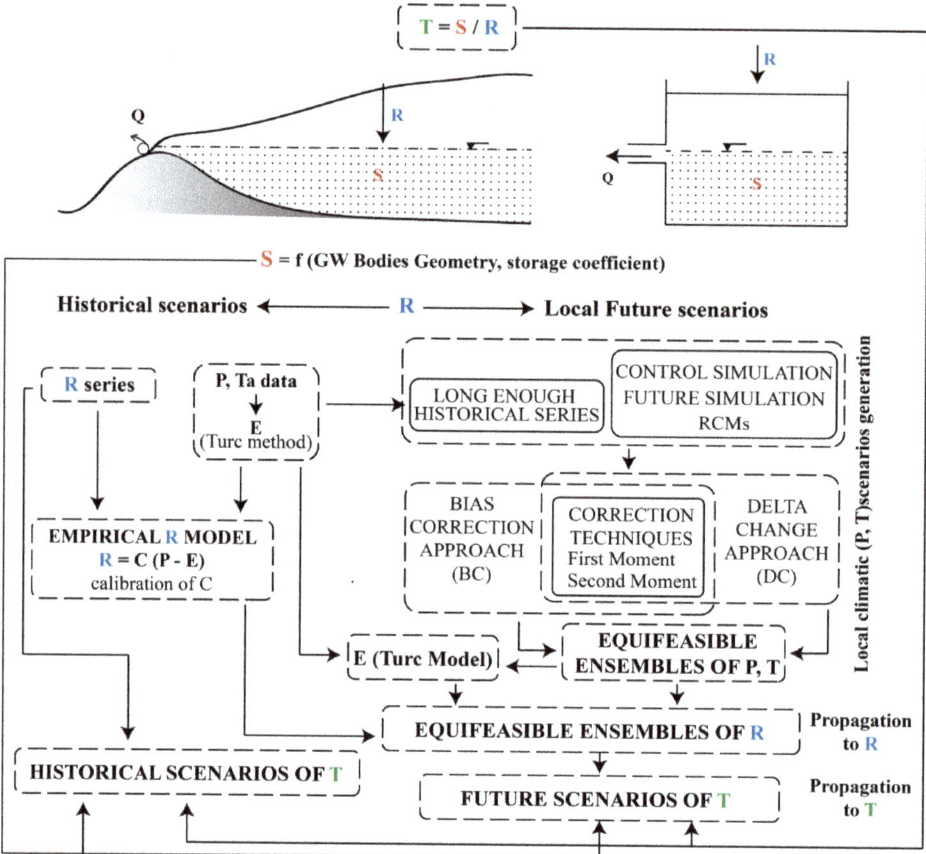

Figure 1. Flowchart of the methodology developed to assess groundwater (GW) bodies' vulnerability to pumping. Notation and units for variables used: P, E, R, and Q are respectively precipitation, actual evapotranspiration, net GW recharge from P, and net GW discharge in mm year^{-1}; Ta is temperature in °C; C and S are respectively a dimensionless effective recharge coefficient (−) and a GW storage (Mm^3); and T is the natural turnover time index in years.

2.2. Materials: Description of the Study Area and the Available Information

2.2.1. Location, Geological Context and Historical Climatic Data

In continental Spain, 717 GW bodies were defined for Water Planning. These GW bodies cover 71% of the territory. The definition of sustainable management strategies for water resource systems should pay special attention to water bodies at risk of not achieving the WFD (2000) objectives. For this reason, in this study we focused on the 146 Spanish GW bodies at risk of not fulfilling the WFD (2000) objectives (see Figure 2) in order to identify which could be potentially considered strategic for a sustainable conjunctive use management of droughts.

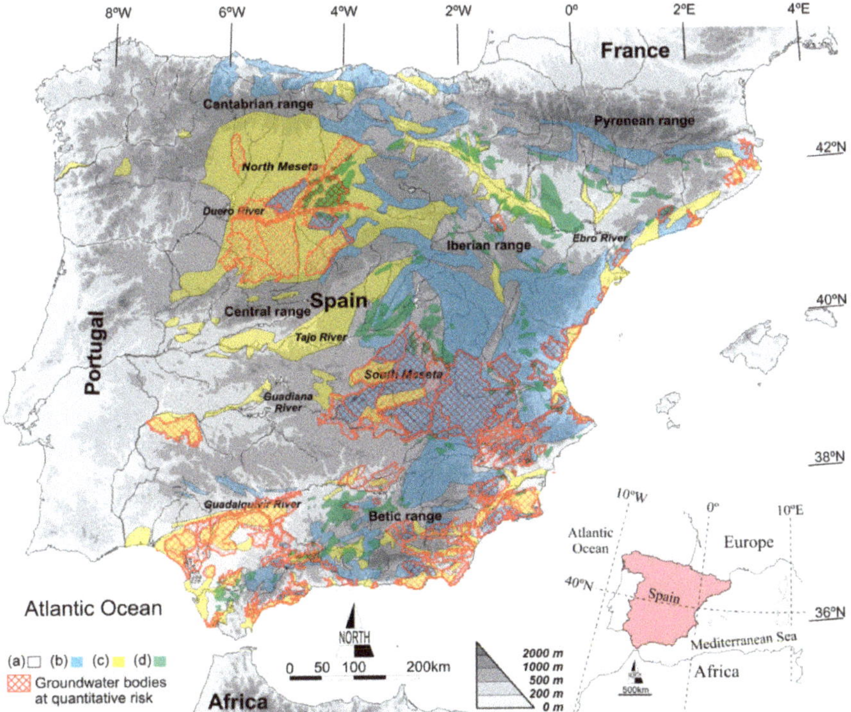

Figure 2. Map of continental Spain, showing the 146 Spanish GW bodies at quantitative risk of not fulfilling the Water Framework Directive (WFD) [14] (2000) objectives (red shadowed areas), the main mountain ranges and hydrographic basins, and the hydrogeological behavior of geological materials forming the GW bodies according to permeability type [37], modified from [31] as: (**a**) low to moderate permeability pre-Triassic metamorphic rocks, granitic outcrops, and Triassic to Miocene marly sedimentary formations; (**b**) moderate to high permeability Paleozoic to Tertiary; (**c**) moderate to high permeability Pleo-Quaternary detritic; and (**d**) Triassic to Miocene evaporitic outcrops.

After the WFD [1] came into effect, the European Environment Agency established guidelines for declaring those GW bodies at risk of not fulfilling a good quantitative and qualitative level in the 2020 horizon, as well as general measures to mitigate negative impacts. Declaration of GW bodies at quantitative risk was based on particular net GW balances resulting from GW extractions and losses (pumping, direct evaporation, net GW discharge, lateral outflows) and the available renewable resources (net GW recharge, irrigation and urban returns, stream losses, and lateral inflows). The exploitation index in each GW body, defined as extractions (pumping) divided by the available renewable resources plus the environmental flow, was proposed as a measure of sustainability; 1 was considered the minimum threshold below which there is a GW imbalance. A GW body is classified as having a bad quantitative status when the exploitation index is above 0.8 and there is a clear piezometric level depletion trend over a large fraction of its surface. These GW bodies cover 16% of continental Spain.

The varied geology of continental Spain determines many relatively small high-yielding GW bodies widely distributed throughout its territory. The most important GW bodies are in Pleo-Quaternary sedimentary formations and Triassic to Tertiary carbonate massifs (Figure 2). The former consists of inland GW bodies surrounded by mountain ranges, small alluvial and piedmont units, and deltaic formations on infilled estuaries in coastal areas. Carbonate massifs are common in quite extensive but compartmentalized areas along the northern, eastern, and southern ranges [37]. To a minor extent, the weathered and fissured granite and Paleozoic shale formations in northern, southern, and north-eastern ranges, contain small aquifers of local significance not catalogued as GW bodies. Attending to hydrogeological behavior of geological materials forming the GW bodies deduced from the permeability type, Alcalá and Custodio [31] classified the geological materials forming the GW bodies as: (a) low to moderate permeability pre-Triassic metamorphic rocks, granitic outcrops, and Triassic to Miocene marly sedimentary formations forming low productive GW bodies and impervious areas; (b) moderate to high permeability Paleozoic to Tertiary carbonates forming mostly highly productive GW bodies; (c) moderate to high permeability Pleo-Quaternary detrital materials corresponding typically to moderately to highly productive GW bodies; and (d) Triassic to Miocene evaporitic outcrops characterizing areas subjected to potential GW pollution due to natural sources of salinity (Figure 2).

For the purposes of this research, historical climatic (temperature and precipitation) data collected from the Spain02 project [38] for the chosen reference period (1976–2005) were used. This precipitation data includes both rainfall and snowfall. Temperature and precipitation show significant spatial heterogeneity as a result of highly variable climatic conditions. Annual mean P ranges from 190 mm year^{-1} in south-eastern semiarid regions to over 2000 mm year^{-1} in humid northern locations (Figure 3a). Nearly all P occurs between late autumn and winter (November to March), due to the circulation of cold air masses formed over the North Atlantic Ocean, in addition to deep pressure lows that move eastwards, resulting in an influx of air masses over the Subtropical Atlantic Ocean [39]. In late summer and autumn months, humid air masses formed over the western Mediterranean Sea may also generate P in eastern coastal regions of Spain, but these events seldom occur far inland [40]. Annual mean Ta varies between 4.6 °C in mountain ranges to 21.1 °C in low-lying large river basin locations (Figure 3b); the coldest and hottest months are January and August, respectively. In any given year, the daily Ta amplitude recorded in southern plateau and river valleys may achieve highs of 50 °C. The pronounced negative gradient of Ta in mountain ranges provides optimum conditions for producing seasonal snow-melt, credited as a principal source of freshwater for filling surface and GW bodies [41].

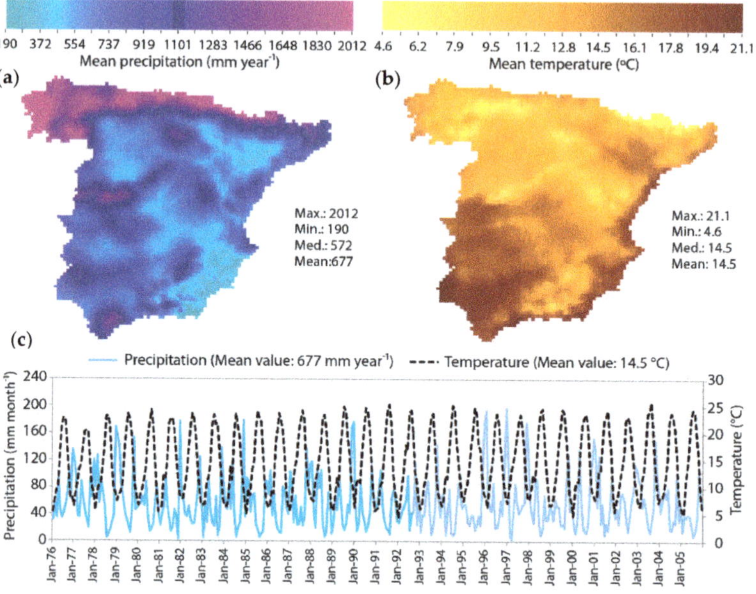

Figure 3. Map of historical mean (**a**) precipitation (mm year^{-1}) and (**b**) temperature (°C) across continental Spain during the reference period (1976–2005), (**c**) temporal series of mean precipitation (mm year^{-1}) Modified from [29].

2.2.2. Estimated Future Climatic Data

In order to generate future local scenarios, the historical climatic data (P and Ta series) in the reference period (1976–2005) were combined with the Climatic model simulations for the Control period (1976–2005) and future scenarios (2016–2045). In this study, we used data generated in previous works to assess the impacts of climate change on R in continental Spain [29]. This includes various climatic model simulations undertaken by the CORDEX EU project [42] for the most pessimistic IPCC emission scenario, the Representative Concentration Pathways 8.5 (RCP8.5). Selected simulations consist of results from five Regional Climate Models (RCMs) (CCLM4-8-17, RCA4, HIRHAM5, RACMO22E, and WRF331F) nested within four distinctive General Circulation Models. An equi-feasible ensemble of all RCM simulations was performed using 1976–2005 as the control/historical reference period and fixing the future horizon scenario as 2016–2045.

The RCPs are the greenhouse gas concentration trajectories adopted by the IPCC. They are named according to the radiative forcing that they represent. Radiative forcing is the change in the net downward minus upward radiative flux at the troposphere or top of the atmosphere due to a change in an external driver of climate change. The RCP8.5 is the most pessimistic pathway for which radiative forcing reaches values greater than 8.5 W m−2 by 2100. The selected RCM projections were performed using simulations of the RCP8.5 trajectories to generate potential future series of P and T. In this work, we corrected these series to generate local scenarios and to propagate their impacts on R.

The RCM climate modelling simulates climate conditions defined with some initial conditions, time-dependent lateral meteorological conditions and surface boundary conditions, to drive high-resolution models. These conditions are typically wind components, temperature, water vapor, and surface pressure. The driving data are derived from GCMs that simulate with a coarse resolution. Table 1 shows the GCMs used by the RCMs employed in this work. The World Climate Research Programme (WCRP) through the CORDEX project guarantees the quality of the RCMs they collected. However, uncertainties related to RCMs can be important and they must be adapted to the study area.

Table 1. Regional Climatic Models (RCMs) and General Circulation Models (GCMs) considered to define the climatic scenarios.

RCMs \ GCMs	CNRM-CM5	EC-EARTH	MPI-ESM-LR	IPSL-CM5A-MR
CCLM4-8-17	X	X	X	
RCA4	X	X	X	
HIRHAM5		X		
RACMO22E		X		
WRF331F				X

The monthly bias of the model within the reference period (1976–2005) was estimated as the mean relative differences between the control simulation and the historical P and Ta time series calculated for each month of an average year. This was used to generate the future series by applying a bias correction technique (scenario E_B). The monthly delta changes between control and future P (2016–2045) were also estimated to generate series by applying a delta change approach (scenario E_D) (Figure 4). Figure 4 shows that the potential future mean P and Ta generated for the scenarios E_B and E_D are exactly the same, although the temporal evolution of these variables is different, due to the different way in which they are generated from the historical values.

2.2.3. Potential GW Storage under the Surface Connection

For each of the selected GW bodies, we have taken the available information about potential GW storage under the surface connection (Figure 5). These were collected from the last River Basin Plans (2015–2021) published by the different River Basin authorities. This summarizes geological and topographical information to define the GW body geometry, that combined with the storage coefficients provide the S (Mm^3) value for these GW bodies.

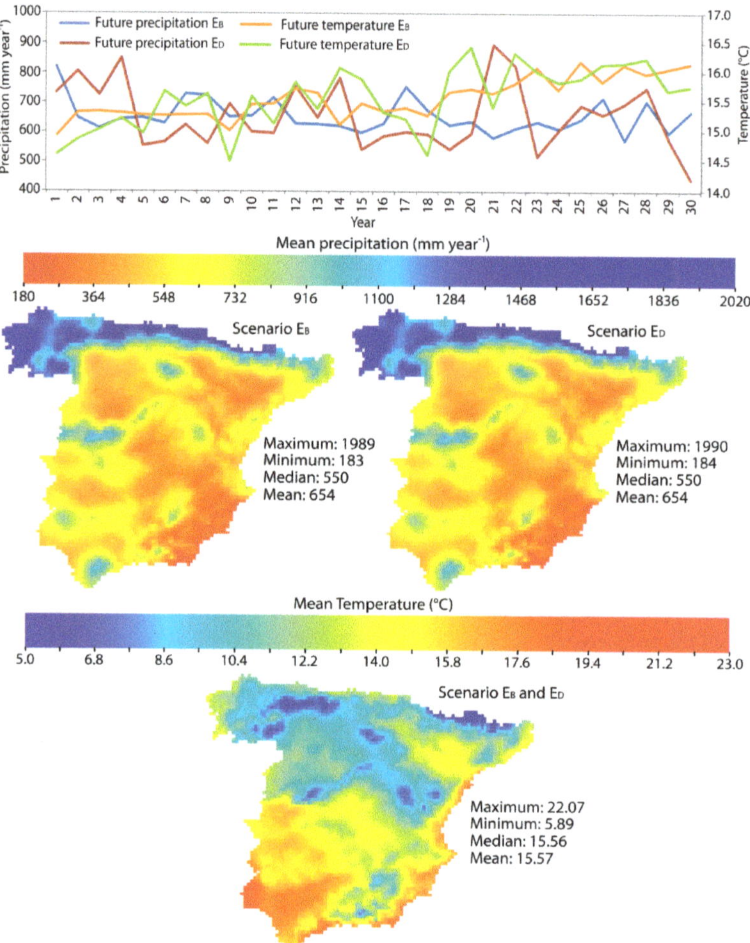

Figure 4. Potential future mean precipitation (mm year^{-1}) and temperature (°C) obtained with the equi-feasible delta and bias ensembles scenarios (E$_D$, E$_B$). Modified from [29].

2.2.4. Net GW Recharge: Historical and Future Scenarios

An empirical precipitation-R model was employed to estimate the historical R within the reference period and the impacts of potential future climatic scenarios on R [29]. It is defined as follows:

$$R = C(P - E) \tag{2}$$

where R, P, and E in mm year^{-1} and dimensionless C are defined in caption of Figure 1. For estimating E, we used the non-global Turc [43,44] formulation:

$$E = \frac{P}{\sqrt{0.9 + \frac{P^2}{L^2}}} \tag{3}$$

where $L = 300 + 25Ta + 0.05Ta^3$ is a dimensionless form parameter of annual temperature.

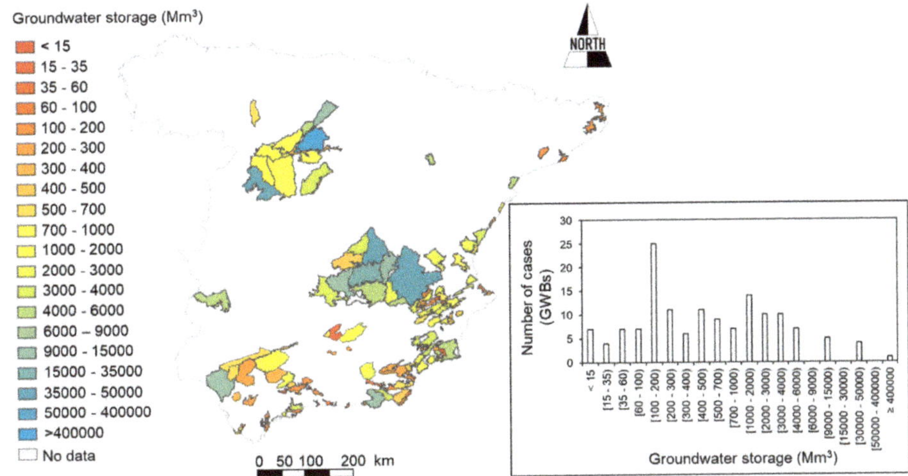

Figure 5. Potential storage capacity (Mm^3) under the surface connection for the 146 Spanish GW bodies at quantitative risk of not achieving the WFD [1] objectives.

This model has been used to propagate the impacts of local historical and future climatic fields in continental Spain (Figure 6).

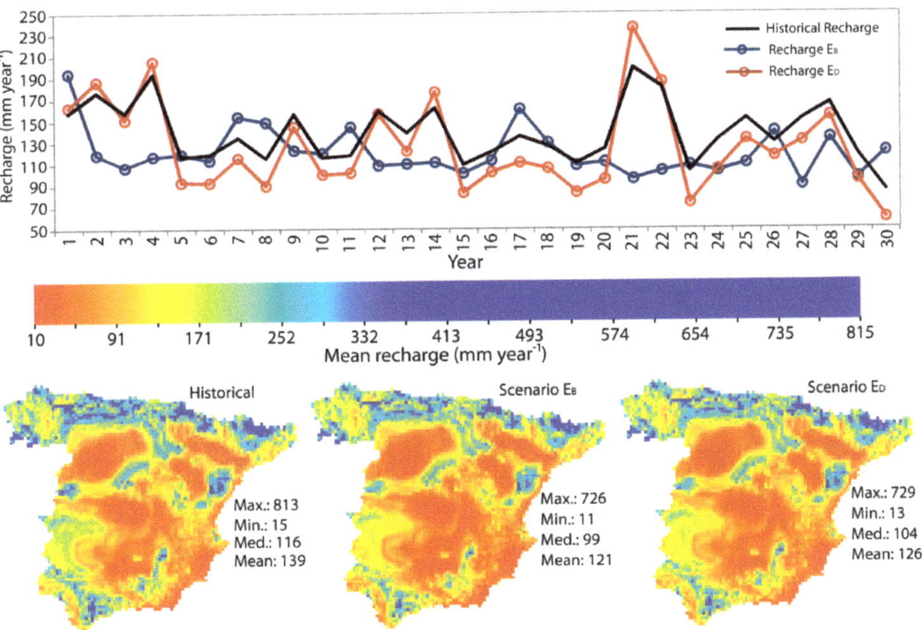

Figure 6. Historical (1976–2005) and future (2011–2045) potential R (mm year^{-1}) for the 2 defined equi-feasible ensemble scenarios. Modified from [29].

3. Results and Discussion

3.1. The T Index in Continental Spain: Historical and Future Scenarios

The information summarized in the previous section was used to assess the natural T for the historical period (reference period 1976–2005) and future potential scenarios in the horizon 2016–2045 that correspond to the RCP 8.5 emission scenario (Figure 7). Two different local climatic scenarios have been considered to assess the potential impacts on T values, one generated by an ensemble of bias correction approaches (E_B) and another by an ensemble of delta change approaches (E_D). The methodology and the series generated for those scenarios were described in Section 2.2.2. Figure 7 shows a heterogeneous distribution of T values within the 146 selected GW bodies as case studies. The box whiskers plot also reflects this wide range of T values moving from a minimum of 0.25 to a maximum of 3693 years in the historical period. The minimum and maximum values in the future scenarios are 0.32 and 4176 years for E_B, and 0.28 and 3953 years for E_D. In order to understand the variability of this value, we should take into account the formulation applied to estimate T (Equation (1)) in each GW body, defined as S divided by R, where S depends on the geometry ("the size" of the GW body) and the storage coefficients (hydrodynamic parameter depending on the geology and hydraulic behavior of the aquifer). Therefore, this variability in T values is logical taking into account the varied geology, size and hydraulic behavior of the considered GW bodies, as shown in Figure 2. The influence of S, and, therefore, the combined influence of geology, size and hydraulic behavior on T, is described in Section 3.2, where a sensitivity analyses of T values to S and R is performed. In addition, we also analyze the influence of the environmental conditions, taking into account that R depends on climatic and ground characteristics.

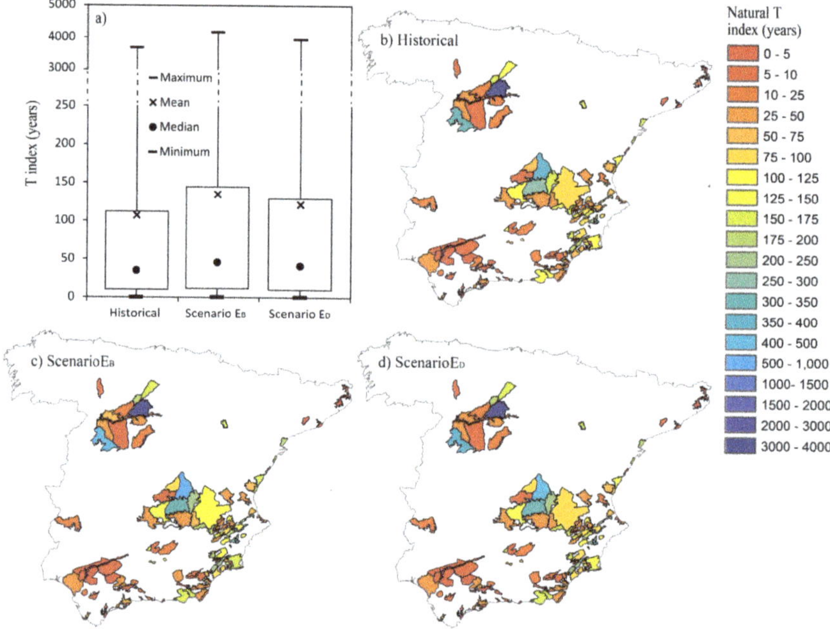

Figure 7. Box-whiskers (**a**) and maps of the T index in the 146 Spanish GW bodies at risk [1]. Historical (**b**) values and future potential scenarios (EB (**c**) and ED (**d**)) in the horizon 2011–2045. The differences between the future scenarios (E_B and E_D) in terms of impacts on the T index are small, due to the differences between the impacts on mean R also being small (see maps of Figure 6). The mean values of R for both scenarios are very similar, although the monthly series are different (see temporal series of Figure 6).

Low T values means that R is close to S, and therefore, they are extremely vulnerable to exploitation, even in periods when pumping is smaller than the average R. This can be especially relevant in areas with scarce resources where long and intensive drought appear and will be exacerbated in the future due to climate change. If we assume that the long term management of the Water Resource Systems allows to maintain the natural mean reserves (the mean S) of the GW bodies, the highest values of T correspond to GW bodies that can be very useful due to their buffer values role in managing drought periods. Around 26.9% of the studied GW bodies show low pumping vulnerability with historical T values above 100 years, with this percentage increasing to 33.1% in the near future horizon values (until 2045).

Taking into account the formulation employed to assess T as S divided by R (see Equation (1)), the impacts of the future scenarios on T are explained by the change in R, which is the only variable that depends on the climatic conditions. The T values will increase in the future in most of the GW bodies (Figure 8) due to the recharge (R) being reduced; meanwhile the total potential storage under the surface connection (S) will stay invariant. The impacts of potential future scenarios on T values will be heterogeneous (see maps of Figure 8). The box whiskers plot also reflects a wide range of T value changes with respect to the historical values moving from a reduction of 2.8 years to an increment of 483 years, which is due also to the variability observed for the recharge, where we estimate changes with respect to the historical between a reduction 47.0 mm year^{-1} and an increment of 2.7 mm year^{-1}.

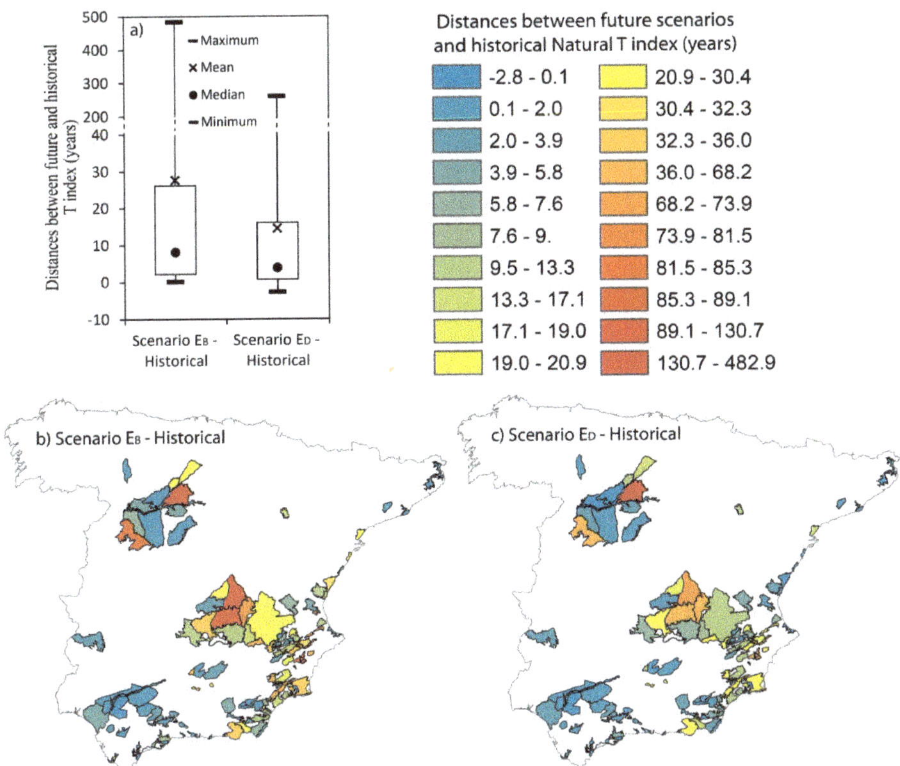

Figure 8. Box-Whiskers (**a**) and maps (**b**,**c**) of the distances between historical natural T and future potential values in horizon 2011–2045.

The increments in T values will force the application of more restrictive long-term management strategies within the systems to maintain the natural mean reserves, but if this long term constraint is fulfilled, the potentiality of those GW bodies to be used to play a buffer role to manage drought periods will be in many cases even higher than in the historical period (Figure 8).

3.2. Influence S and R on the T Index for Different Lythologies

The T–S and T–R data-pairs were compared attending to the predominant lithology of each GW body at risk. In general, for a given lithology, higher correlation is observed for T–S values (Figure 9a) than for T–R (Figure 9b). This can be observed more clearly in carbonated GW bodies ($R^2 = 0.82$ for T–S relationship and 0.04 for T–R), while in the detrital GW bodies R^2 would be 0.47 and 0.15, respectively. This could be explained by the lower variation of R in the carbonated GW bodies (see Figure 9). Nevertheless, in general the variability of the R values (ranging between 22.8 and 309.7 mm year^{-1}) is significantly smaller than the S values (ranging between 2.9 and 401,1875 Mm^3), which is the variable that better explains the high variability of T values (ranging between 0.3 and 3693.2 years), especially in carbonated GW bodies, where the variability of R is very low.

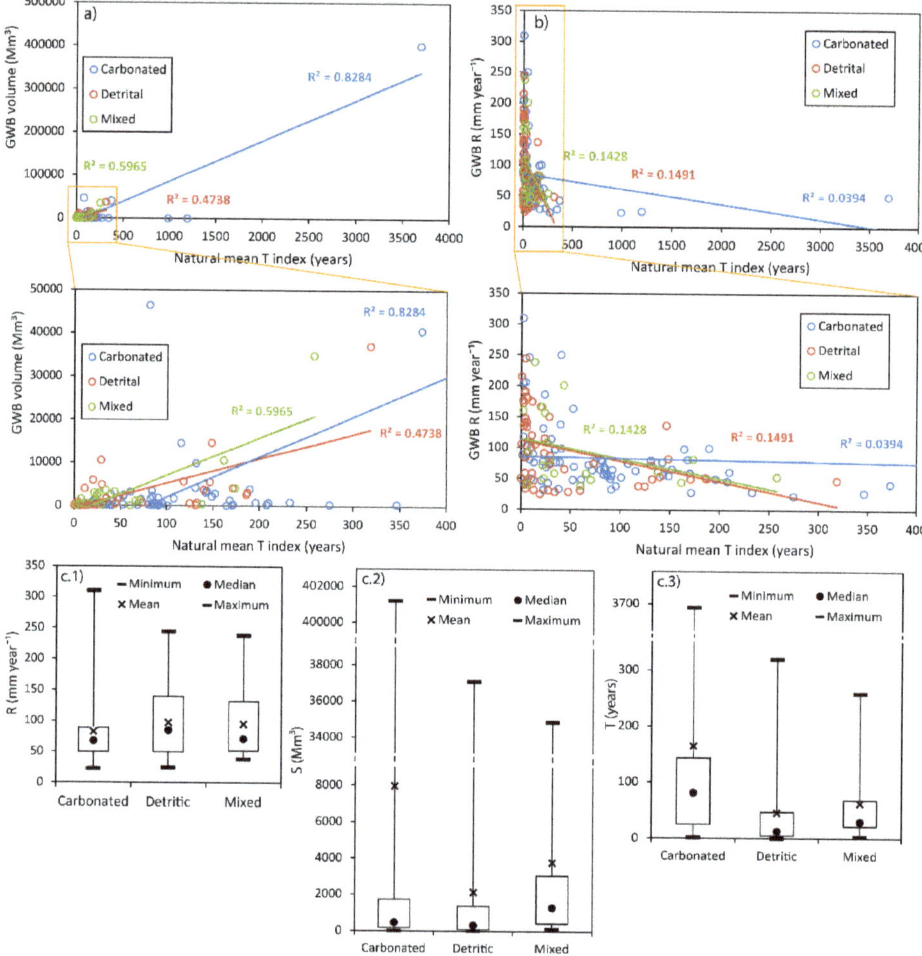

Figure 9. (**a**) GW body volume (Mm^3) vs. natural mean T index (year), (**b**) R (mm year^{-1}) vs. natural mean T index (year), T and (**c**) box-whiskers of R, S and T for the three considered GW bodies lithological categories: carbonated, detrital, and mixed.

This analysis has also been extended to other explanatory variables, such as the "recharge coefficients" defined as RC = R/P (See Figure 10b); and the effective recharge coefficients C = R/(P-E) (see Figure 10c and Equation (2)). This work assumes that C is a parameter that will stay invariant when assessing future impacts on GW bodies R, which is a common assumption performed in future projection studies [35,45], despite that, theoretically, they may vary due to changes in vegetation cover, land use, soil properties, and structure of rainfall events, as foreseen by the Spanish desertification model prepared in the framework of the National Plan to Combat Desertification [46,47].

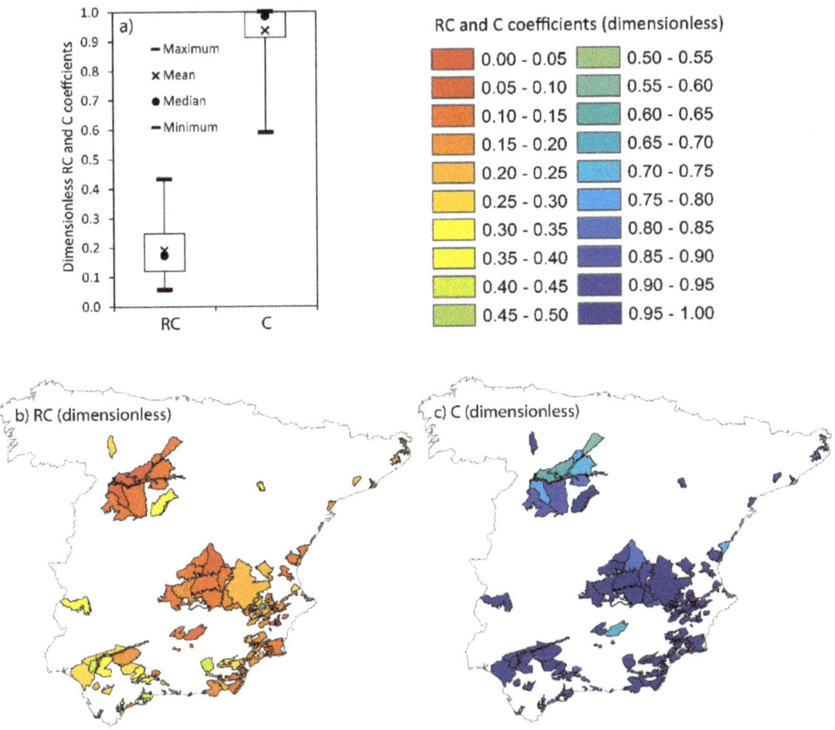

Figure 10. Box-Whisker (**a**) and maps of the of the "recharge coefficients" (RC) (**b**) and the "effective recharge coefficient (C)" (**c**) in the GW bodies at risk [1].

The difference between C and RC gives an idea of the impacts of the actual evapotranspiration in the calculation of recharge. Higher differences indicate longer distance between precipitation and effective precipitation, defined as the precipitation minus the actual evapotranspiration. We found that the T changes (future scenarios vs. historical period) are higher when the difference between C and RC are higher (see Figure 11). Therefore, the GW bodies with higher difference between C and RC are more sensitive to climate change.

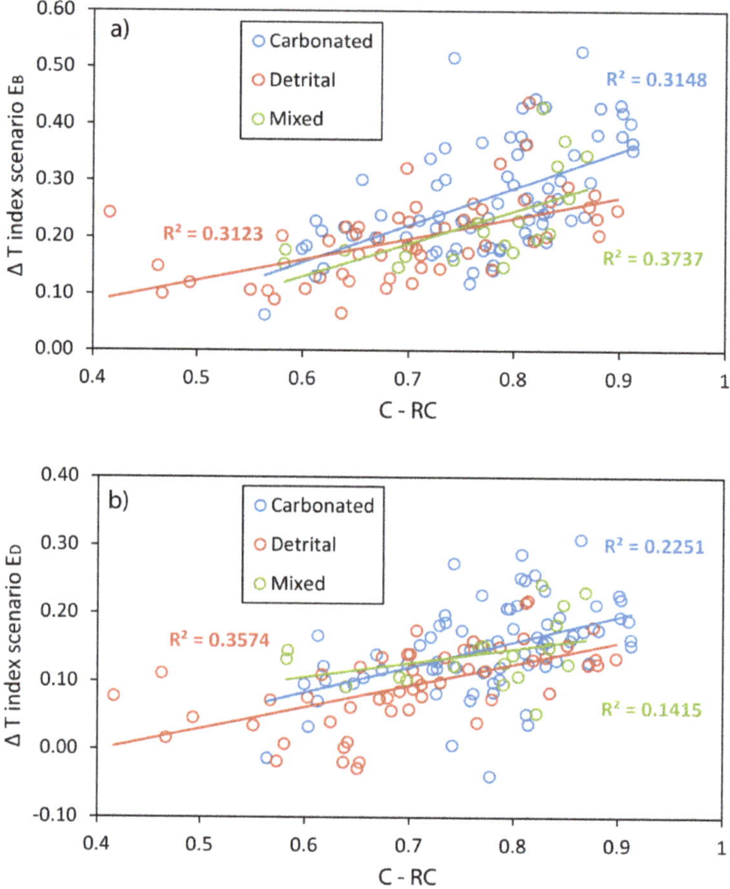

Figure 11. Absolute distance (years) of future T values (**a**) EB scenario and (**b**) ED scenario, with respect to the historical T vs. difference between effective recharge coefficients and recharge coefficients.

3.3. Hypothesis Assumed and Limitations of the Method (Recharge and Total Storage Uncertainty)

In order to estimate the natural mean GW renewable time, we assume that there are no changes in Land Use and Land Cover (LULC), no pumping, and the net GW discharge will start when the potential total GW storage reaches the level of the surface connection (spring, streambed, river level, or sea level boundary condition) (See Figure 1). As described, a parallelism between unconfined aquifers and reservoirs is adopted to approach this by the T index. Analogies between an unconfined aquifer and a reservoir have been previously adopted to approach different stream-aquifer interaction problems [48–51] The T values are used to assess vulnerability to pumping during droughts, assuming that the LULC and the management of the water resources system will allow the natural mean reserves (the mean S) of the GW bodies to be maintained, even under long term future R scenarios. Under this assumption, highest values of T correspond to GW bodies that can be useful to manage droughts due to their buffer values role.

We assume that the historical S values derived from the last River Basin Plans published by the different River Basin authorities (2015–2021) are good enough for a preliminary approximation of the T values.

The historical R in the reference period (1976–2005) has been estimated by applying an empirical approach [29] with a spatial resolution of 10 km × 10 km, which is considered an accurate enough approach for a preliminary simple assessment of T. The R model is based on a preliminary approach to the main drivers of the R dynamic (precipitation, actual evapotranspiration and "effective recharge coefficients") in accordance with the available data and, although it is not a state of the art model, it produces a good enough preliminary approach to identify potential strategic GW bodies, where more detailed studies will be required for a more accurate assessment. The main hypothesis and limitations of the proposed approach are:

- The climatic fields (P and Ta) in the case studies are approximated by the Spain02 project dataset [38]. This dataset has been recently validated by Quintana-Seguí et al. [52] and has already been employed in many research studies.
- We assume that the mean yearly long term E assessment provided by the non-global Turc's model [43,44], whose results depend on mean annual Ta and P, is good enough for this preliminary assessment. Due to its simplicity and efficiency this approach has been extensively applied in research works in which preliminary E assessment for historical and/or future scenarios is included [33,53]. In spite of this, more accurate assessments of groundwater resource will require, in addition to the non-global E approach, corrections by using global models for E [54] or an external calibration by using well-suited recharge functions [55]. This is especially interesting in the Spanish drylands, where E is typically close to P [53,54]. In this work we use the second option, i.e., the Turc formulation. and the "effective recharge coefficients" deduced from a previous calibrated recharge function provided by Alcalá and Custodio [31,32].
- The "effective recharge coefficients" (C) are calibrated from the R values derived by Alcalá and Custodio, [31,32]. They used a chloride mass balance method whose accuracy is similar to that obtained when global models for E are used for recharge purposes [55]. These values are available at a spatial resolution of 10 km x 10 km grid, and therefore the R model will also be at this scale. We assume that the future impacts of potential climatic scenarios on R can be obtained by propagating future local scenarios with the R model previously calibrated. Therefore, this assumes that the effective recharge coefficient remains invariant in simulating future conditions.
- The analyses of future potential climatic scenarios do not include the simulation of any future LULC scenarios and/or management scenarios of the water resources system. We only analyze potential impacts on T, and, therefore, on R, due to climate drivers, assuming that, in the future, the LULC and management will allow the maintenance of long term mean natural reserves. Under this assumption, the proposed method will be useful in the future to identify strategic resources to manage droughts. In the literature we found several research works in which the potential future impacts on aquifers are analyzed taking into account only climate drivers (Pulido-Velazquez et al., 2018; Pardo-Iguzquiza et al., 2019). The development of additional research works will be needed to study specific LULC and management issues (Pulido-Velazquez et al., 2018).

Future local climate driven scenarios have been generated for a short-term horizon (2015–2045) assuming the most pessimistic emission scenarios RCP8.5.

- Local projections have been obtained from different climatic model simulations by applying two downscaling approaches (correction of first and second order moments) under two different hypotheses (bias correction and delta change techniques) [33,35].
- The final scenarios employed to study potential impacts on R and T have been defined by an equi-feasible ensemble of local projections, which produce more robust and representative projections than those based on a single model [36].

4. Conclusions

Aquifers with higher GW mean residence time show lower vulnerability to pumping during drought periods. T values are used to assess vulnerability to pumping during drought periods, assuming that the long-term management of the Water Resources Systems will allow for the maintenance of natural mean reserves (the mean S) of the GW bodies, even under future long-term recharge (R) scenarios. A preliminary assessment of this variable can be obtained by the natural mean turnover time (T) index defined as S divided by R. Aquifers where R is close to S are extremely vulnerable to exploitation, even in periods when pumping is smaller than the average R. This can be especially relevant in areas with scarce resources where long and intensive droughts appear and will be exacerbated in the future due to climate change. In this work we identify potential strategic GW resources, with low vulnerability to pumping, which can be useful to define sustainable conjunctive use management of droughts in continental Spain. We focus our analyses on the Spanish GW bodies at risk of not achieving the European Water Framework Directive [14] objectives (146 GW bodies). We performed a historical and future (short term period until 2045) assessment of T as the S/R ratio. Around 26.9% of these GW bodies show low pumping vulnerability with historical T values above 100 years, with this percentage increasing to 33.1% in the near future horizon values (until 2045). The results observed in the study area show a significant heterogeneity. The maximum range of the historical T variability is around 3700 years, which also increases in the near future to 4200 years. Therefore, the vulnerability to pumping is also quite heterogeneous. The T index values will change in future horizons, and the potential use and the constraints to be applied in using GW bodies to define conjunctive use strategies in order to adapt to Climate Change scenarios will also change in the coming years. We have also analyzed the variability and influence of R and S values in the determination of T for different aquifer lithologies.

Author Contributions: D.P.-V. planned the research/methodology, and contribute to writing and reviewing the manuscript. J.R.: contribution to the assessments, figures and writing activities. A.-J.C.-L.: contribution to the assessments, figures and writing activities. F.J.A.: contribution to writing and reviewing the manuscript. F.F.-C.: contribution to the assessments. L.B.-R.: contribution in figures and writing activities. Valuable comments and suggestions were provided by four anonymous referees. All authors have read and agreed to the published version of the manuscript.

Funding: This research has been partially funded by the projects GeoE.171.008-TACTIC and GeoE.171.008-HOVER, funded by European Union's Horizon 2020 research and innovation program; and SIGLO-AN (RTI2018-101397 -B-I00) project from the Spanish Ministry of Science, Innovation and Universities (Programa Estatal de I+D+I orientada a los Retos de la Sociedad).

Acknowledgments: We thanks the reviewers and the editor for the valuable comments they provided to improve the manuscript.

Conflicts of Interest: The authors declare that they have no known competing financial interests or personal relationships that could have appeared to influence the work reported in this paper.

Abbreviations

S	Groundwater storage
R	Net groundwater recharge
T	Natural mean groundwater turnover time
GW	Groundwater
WFD	European Water Framework Directive
E	Actual evapotranspiration
P	Precipitation
Ta	Temperature
Q	Net groundwater discharge
U	Groundwater pumping
RC	Recharge coefficient or P-to-R ratio
C	Effective recharge coefficient

References

1. Hayes, M.J.; Svoboda, M.; Wall, N.A.; Widhalm, M. The Lincoln Declaration on Drought Indices: Universal Meteorological Drought Index Recommended. *Bull. Am. Meteorol. Soc.* **2011**, *92*, 485–488. [CrossRef]
2. Mishra, A.K.; Singh, V.P. A review of drought concepts. *J. Hydrol.* **2010**, *391*, 202–216. [CrossRef]
3. Mehran, A.; Mazdiyasni, O.; AghaKouchak, A. A hybrid framework for assessing socioeconomic drought: Linking climate variability, local resilience, and demand. *J. Geophys. Res. Atmos.* **2015**, *120*, 7520–7533. [CrossRef]
4. Blenkinsop, S.; Fowler, H.J. Changes in drought characteristics for Europe projected by the PRUDENCE regional climate models. *Int. J. Climatol.* **2007**, *27*, 1595–1610. [CrossRef]
5. Collados-Lara, A.-J.; Pulido-Velazquez, D.; Pardo-Igúzquiza, E. An Integrated Statistical Method to Generate Potential Future Climate Scenarios to Analyse Droughts. *Water* **2018**, *10*, 1224. [CrossRef]
6. LeDuc, C.; Pulido-Bosch, A.; Remini, B. Anthropization of groundwater resources in the Mediterranean region: Processes and challenges. *Hydrogeol. J.* **2017**, *25*, 1529–1547. [CrossRef]
7. Baena-Ruiz, L.; Pulido-Velazquez, D. A Novel Approach to Harmonize Vulnerability Assessment in Carbonate and Detrital Aquifers at Basin Scale. *Water* **2020**, *12*, 2971. [CrossRef]
8. Pulido-Velazquez, D.; García-Aróstegui, J.L.; Molina, J.-L.; Pulido-Velazquez, M. Assessment of future groundwater recharge in semi-arid regions under climate change scenarios (Serral-Salinas aquifer, SE Spain). Could increased rainfall variability increase the recharge rate? *Hydrol. Process.* **2014**, *29*, 828–844. [CrossRef]
9. Tramblay, Y.; Koutroulis, A.; Samaniego, L.; Vicente-Serrano, S.M.; Volaire, F.; Boone, A.; Le Page, M.; Llasat, M.C.; Albergel, C.; Burak, S.; et al. Challenges for drought assessment in the Mediterranean region under future climate scenarios. *Earth-Sci. Rev.* **2020**, *210*, 103348. [CrossRef]
10. Besbes, M.; Chahed, J.; Hamdane, A. *Sécurité Hydrique de la Tunisie: Gérer l'Eau en Conditions de Pénurie [Water Security of Tunisia: Managing Water in Scarcity Conditions]*; L'Harmattan: Paris, France, 2014.
11. Custodio, E.; Andreu-Rodes, J.M.; Aragón, R.; Estrela, T.; Ferrer, J.; García-Aróstegui, J.L.; Manzano, M.; Rodríguez-Hernández, L.; Sahuquillo, A.; Del Villar, A. Groundwater intensive use and mining in south-eastern peninsular Spain: Hydrogeological, economic and social aspects. *Sci. Total. Environ.* **2016**, *559*, 302–316. [CrossRef]
12. Werner, A.D.; Bakker, M.; Post, V.E.; Vandenbohede, A.; Lu, C.; Ataie-Ashtiani, B.; Simmons, C.T.; Barry, D. Seawater intrusion processes, investigation and management: Recent advances and future challenges. *Adv. Water Resour.* **2013**, *51*, 3–26. [CrossRef]
13. Llamas, M.R.; Garrido, A. Lessons from intensive groundwater use in Spain: Economic and social benefits and conflicts. In *The Agricultural Groundwater Revolution: Opportunities and Threats to Development*; CABI Publishing: Wallingford, UK, 2009; pp. 266–295.
14. Water Framework Directive (WFD). *Directiva 2000/60/CE del Parlamento Europeo y del Consejo de 23 de Octubre de 2000. Diario Oficial de las Comunidades Europeas de 2212/2000. L 327/1–327/32*; Water Framework Directive: Brussel, Belgium, 2000.
15. Machiwal, D.; Cloutier, V.; Güler, C.; Kazakis, N. A review of GIS-integrated statistical techniques for groundwater quality evaluation and protection. *Environ. Earth Sci.* **2018**, *77*, 681. [CrossRef]
16. Krogulec, E. Intrinsic and Specific Vulnerability of Groundwater in a River Valley—Assessment, Verification and Analysis of Uncertainty. *J. Earth Sci. Clim. Chang.* **2013**, *4*. [CrossRef]
17. Gogu, R.C.; Dassargues, A. Current trends and future challenges in groundwater vulnerability assessment using overlay and index methods. *Environ. Earth Sci.* **2000**, *39*, 549–559. [CrossRef]
18. Focazio, M.J.; Reilly, T.E.; Rupert, M.G.; Helsel, D.R. Assessing ground-water vulnerability to contamination: Providing scientifically defensible information for decision makers. In *Circular*; US Geological Survey: Reston, VA, USA, 2002.
19. Aller, L.; Bennet, T.; Lehr, J.H.; Petty, R.J.; Hackett, G. *DRASTIC: A Standardized System for Evaluating Groundwater Pollution Potential Using Hydrogeological Settings*; EPA/600/2- 543 87/035; US Environmental Protection Agency: Ada, OK, USA, 1987; p. 641.
20. Van Stempvoort, D.; Ewert, L.; Wassenaar, L. Aquifer vulnerability index: A GIS 643 Compatible method for groundwater vulnerability mapping. *Can. Water Resour. J.* **1993**, *18*, 25–37. [CrossRef]
21. Civita, M.; De Maio, M. Assessing and mapping groundwater vulnerability to contamination: The Italian "combined" approach. *Geofis. Int.* **2004**, *43*, 513–532. [CrossRef]

22. Busico, G.; Kazakis, N.; Cuoco, E.; Colombani, N.; Tedesco, D.; Voudouris, K.; Mastrocicco, M. A novel hybrid method of specific vulnerability to anthropogenic pollution using multivariate statistical and regression analyses. *Water Res.* **2020**, *171*. [CrossRef]

23. Maloszewski, P.; Zuber, A. Determinig the turnover time of GROUNDWATER systems with the aid of environmental tracers, I. Models and their applicability. *J. Hydrol.* **1982**, *57*, 207–231. [CrossRef]

24. Chambers, L.A.; Gooddy, D.C.; Binley, A.M. Use and application of CFC-11, CFC-12, CFC-113 and SF6 as environmental tracers of GROUNDWATER residence time: A review. *Geosci Front.* **2018**, *10*, 1643–1652. [CrossRef]

25. Sanford, W.E. Calibration of models using groundwater age. *Hydrogeol. J.* **2011**, *19*, 13–16. [CrossRef]

26. Jurgens, B.; Böhlke, J.K.; Kauffman, L.; Belitz, K.; Esser, B. A partial exponential lumped parameter model to evaluate GROUNDWATER age distributions and nitrate trends in long-screened wells. *J. Hydrol.* **2016**, *543*, 109–126. [CrossRef]

27. Eberts, S.M.; Böhlke, J.K.; Kauffman, L.J.; Jurgens, B.C. Comparison of particle-tracking and lumped-parameter age-distribution models for evaluating vulnerability of production wells to contamination. *Hydrogeol. J.* **2012**, *20*, 263–282. [CrossRef]

28. Skaugen, T.; Astrup, M.; Roald, L.A.; Førland, E. Scenarios of extreme daily precipitation for Norway underclimate change. *Hydrol. Res.* **2004**, *35*, 1–13. [CrossRef]

29. Pulido-Velazquez, D.; Renau-Pruñonosa, A.; Llopis-Albert, C.; Morell, I.; Collados-Lara, A.J.; Senent-Aparicio, J.; Baena-Ruiz, L. Integrated assessment of future potential global change scenarios and their hydrological impacts in coastal aquifers. A new tool to analyse management alternatives in the Plana Oropesa-Torreblanca aquifer. *Hydrol. Earth Syst. Sci.* **2018**, *22*, 3053–3074. [CrossRef]

30. Pulido-Velazquez, D.; Collados-Lara, A.-J.; Alcalá, F.J. Assessing impacts of future potential climate change scenarios on aquifer recharge in continental Spain. *J. Hydrol.* **2018**, *567*, 803–819. [CrossRef]

31. Alcalá, F.J.; Custodio, E. Spatial average aquifer recharge through atmospheric chloride mass balance and its uncertainty in continental Spain. *Hydrol. Process.* **2014**, *28*, 218–236. [CrossRef]

32. Alcalá, F.J.; Custodio, E. Natural uncertainty of spatial average aquifer recharge through atmospheric chloride mass balance in continental Spain. *J. Hydrol.* **2015**, *524*, 642–661. [CrossRef]

33. Collados-Lara, A.-J.; Pulido-Velazquez, D.; Pardo-Igúzquiza, E. A Statistical Tool to Generate Potential Future Climate Scenarios for Hydrology Applications. *Sci. Program.* **2020**, *2020*, 1–11. [CrossRef]

34. Sunyer, M.A.; Hundecha, Y.; Lawrence, D.; Madsen, H.; Willems, P.; Martinkova, M.; Vormoor, K.; Bürger, G.; Hanel, M.; Kriaučiuniene, J.; et al. Inter-comparison of statistical downscaling methods for projection of extreme precipitation in Europe. *Hydrol. Earth Syst. Sci.* **2015**, *19*, 1827–1847. [CrossRef]

35. Collados-Lara, A.J.; Pulido-Velazquez, D.; Mateos, R.M.; Ezquerro, P. Potential Impacts of Future Climate Change Scenarios on Ground Subsidence. *Water* **2020**, *12*, 219. [CrossRef]

36. Agencia Estatal de Meteorología (AEMET). *Generación de Escenarios Regionalizados de Cambio Climático para España*; Agencia Estatal de Meteorología, Mto, Medio Ambiente, Medio Rural y Marino: Madrid, Spain, 2009.

37. IGME. *Groundwater in Spain: Synthetic Study*; ISBN 84-7840-039-7. Technical Report; Ministry of Industry and Energy of Spain: Madrid, Spain, 1993; p. 591. (In Spanish)

38. Herrera, S.; Fernández, J.; Gutiérrez, J.M. Update of the Spain02 Gridded Observational Dataset for Euro-CORDEX evaluation: Assessing the Effect of the Interpolation Methodology. *Int. J. Climatol.* **2016**, *36*, 900–908. [CrossRef]

39. Trigo, R.; Pozo-Vazquez, D.; Osborn, T.; Castro-Díez, Y.; Gámiz-Fortis, S.; Esteban-Parra, M.-J. North Atlantic Oscillation Influence on Precipitation, River Flow and Water Resources in the Iberian Peninsula. *Int. J. Climatol.* **2004**, *24*, 925–944. [CrossRef]

40. Martín-Vide, J.; López-Bustins, J.A. The Western Mediterranean Oscillation and Rainfall in the Iberian Peninsula. *Int. J. Climatol.* **2006**, *26*, 1455–1475. [CrossRef]

41. MIMAM. *White Book of Eater in Spain*; Ministry of the Environment, State Secretary of Waters and Coast, General Directorate of Hydraulic Works and Water Quality: Madrid, Spain, 2000; p. 637, (In Spanish). Available online: http://www.cedex.es/CEDEX/LANG_CASTELLANO/ORGANISMO/CENTYLAB/CEH/Documentos_Descargas/LB_LibroBlancoAgua.htm (accessed on 20 November 2020).

42. CORDEX PROJECT. The Coordinated Regional Climate Downscaling Experiment CORDEX. Program Sponsored by World Climate Research Program (WCRP). 2013. Available online: http://www.cordex.org/ (accessed on 8 September 2018).

43. Turc, L. Water balance of soils: Relationship between precipitation, evapotranspiration and runoff. *Ann. Agron.* **1954**, *5*, 491–569. (In French)

44. Turc, L. Estimation of irrigation water requirements, potential evapotranspiration: A simple climatic formula evolved up to date. *Ann. Agron.* **1961**, *12*, 13–49. (In French)

45. Pardo-Igúzquiza, E.; Collados-Lara, A.J.; Pulido-Velazquez, D. Potential future impact of climate change on recharge in the Sierra de las Nieves (southern Spain) high-relief karst aquifer using regional climate models and statistical corrections. *Environ. Earth Sci.* **2019**, *78*, 598. [CrossRef]

46. Martínez-Valderrama, J.; Ibáñez, J.; Del Barrio, G.; Sanjuán, M.E.; Alcalá, F.J.; Martínez-Vicente, S.; Ruiz, A.; Puigdefábregas, J. Present and future of desertification in Spain: Implementation of a surveillance system to prevent land degradation. *Sci. Total Environ.* **2016**, *563–564*, 169–178. [CrossRef]

47. Martínez-Valderrama, J.; Ibáñez, J.; Alcalá, F.J.; Martínez, S. SAT: A Software for Assessing the Risk of Desertification in Spain. *Sci. Program.* **2020**, *2020*, 7563928. [CrossRef]

48. Pulido-Velazquez, D.; Sahuquillo, A.; Andreu, J.; Pulido-Velazquez, M. An efficient conceptual model to simulate surface water body-aquifer interaction in Conjunctive Use Management Models. *Water Resour. Res.* **2007**, *43*, W07407. [CrossRef]

49. Pulido-Velazquez, D.; Sahuquillo, A.; Andreu, J.; Pulido-Velazquez, M. A general methodology to simulate GROUNDWATER flow of unconfined aquifers with a reduced computational cost. *J. Hydrol.* **2007**, *338*, 42–56. [CrossRef]

50. Pulido-Velazquez, D.; Sahuquillo, A.; Andreu, J. A two-step explicit solution of the Boussinesq equation for efficient simulation of unconfined aquifers in conjunctive-use models. *Water Resour. Res.* **2006**, *42*, W05423. [CrossRef]

51. Llopis-Albert, C.; Pulido-Velazquez, D. Using MODFLOW code to approach transient hydraulic head with a sharp-interface solution. *Hydrol. Process* **2015**, *29*, 2052–2064. [CrossRef]

52. Quintana-Seguí, P.; Turco, M.; Herrera, S.; Miguez-Macho, G. Validation of a new SAFRAN-based gridded precipitation product for Spain and comparisons to Spain02 and ERA-Interim. *Hydrol. Earth Syst. Sci.* **2017**, *21*, 2187–2201. [CrossRef]

53. Were, A.; Villagarcía, L.; Domingo, F.; Alados-Arboledas, L.; Puigdefábregas, J. Analysis of effective resistance calculation methods and their effect on modelling evapotranspiration in two different patches of vegetation in semi-arid SE Spain. *Hydrol. Earth Syst. Sci.* **2007**, *11*, 1529–1542. [CrossRef]

54. España, S.; Alcalá, F.J.; Vallejos, Á.; Pulido-Bosch, A. A GIS tool for modelling annual diffuse infiltration on a plot scale. *Comput. Geosci.* **2013**, *54*, 318–325. [CrossRef]

55. Alcalá, F.J.; Cantón, Y.; Contreras, S.; Were, A.; Serrano-Ortiz, P.; Puigdefábregas, J.; Solé-Benet, A.; Custodio, E.; Domingo, F. Diffuse and concentrated recharge evaluation using physical and tracer techniques: Results from a semiarid carbonate massif aquifer in southeast Spain. *Environ. Earth Sci.* **2011**, *62*, 541–557. [CrossRef]

Publisher's Note: MDPI stays neutral with regard to jurisdictional claims in published maps and institutional affiliations.

MDPI
St. Alban-Anlage 66
4052 Basel
Switzerland
Tel. +41 61 683 77 34
Fax +41 61 302 89 18
www.mdpi.com

Water Editorial Office
E-mail: water@mdpi.com
www.mdpi.com/journal/water